T0206332

Classical Beam Theories of Structural Mechanics

Andreas Öchsner

Classical Beam Theories
of Structural Mechanics

 Springer

Andreas Öchsner
Faculty of Mechanical Engineering
Esslingen University of Applied Sciences
Esslingen am Neckar, Baden-Württemberg
Germany

ISBN 978-3-030-76037-3   ISBN 978-3-030-76035-9   (eBook)
https://doi.org/10.1007/978-3-030-76035-9

This Springer imprint is published by the registered company Springer Nature Switzerland AG
The registered company address is: Gewerbestrasse 11, 6330 Cham, Switzerland

Preface

The understanding of basic, i.e., one-dimensional structural members, is essential in applied mechanics. A systematic and thorough introduction to the theoretical concepts for one-dimensional members keeps the requirements on engineering mathematics quite low, and allows for a simpler transfer to higher order structural members.

Partial differential equations lay the foundation to mathematically describe the mechanical behavior of all classical structural members known in engineering mechanics. Based on the three basic equations of continuum mechanics, i.e., the kinematics relationship, the constitutive law, and the equilibrium equation, these partial differential equations that describe the physical problem can be derived. Nevertheless, the fundamental knowledge from the first years of engineering education, i.e., higher mathematics, physics, materials science, applied mechanics, design, and programming skills, might be required to master this topic.

This monograph provides a systematic and thorough overview of the classical bending members based on the theory for thin beams (shear-rigid) according to Euler-Bernoulli, and the theories for thick beams (shear-flexible) according to Timoshenko and Levinson. The new approach in this textbook is that single-plane bending in the x-y plane as well in the x-z plane is equivalently treated and finally applied to the case of unsymmetrical bending. The fundamental understanding of these one-dimensional members allows a simpler understanding of thin and thick plate bending members.

Esslingen, Germany
April 2021

Andreas Öchsner

Contents

Symbols and Abbreviations

Latin symbols (capital letters)

A area
A_s shear area
D diameter
$\boldsymbol{D}$ compliance matrix
E Young's modulus
EA tensile stiffness
EI bending stiffness
F force
G shear modulus
GA shear stiffness
$\mathcal{H}$ first moment of area
I second moment of area
$\boldsymbol{K}$ global stiffness matrix
L length
M moment
$\boldsymbol{M}$ mass matrix
N normal force (internal reaction)
Q shear force (internal reaction)
R radius

Latin symbols (small letters)

a geometric dimension, factor
b body force, factor, geometric dimension
$\boldsymbol{b}$ column matrix of distributed loads
c constant of integration, geometric dimension

d geometric dimension
e column matrix of generalized strains
f scalar function
$\boldsymbol{f}$ column matrix of loads
h geometric dimension
k elastic foundation modulus
k_s shear correction factor
m distributed moment
$\boldsymbol{n}$ normal vector
p distributed load in x-direction
q distributed load in y- or z-direction
s geometric dimension
$\boldsymbol{s}$ column matrix of generalized stresses
t thickness, time
$\boldsymbol{t}$ tangential vector
u displacement
$\boldsymbol{u}$ column matrix of deformations
x Cartesian coordinate
y Cartesian coordinate
z Cartesian coordinate

Greek symbols (capital letters)

Γ boundary
Π total strain energy
Φ eigenmode
Ω domain

Greek symbols (small letters)

α rotation angle
β rotation angle
γ shear strain (engineering definition)
ε strain
ζ coordinate
η coordinate
θ function
κ curvature
λ factor
ν Poisson's ratio
σ normal stress

τ shear stress
ϕ rotation (Timoshenko or Levinson beam)
φ rotation (Euler-Bernoulli beam)
ψ function
ω eigenfrequency

Mathematical symbols

$[\ldots]^{\mathrm{T}}$ transpose
$\det(\ldots)$ determinant
$\mathcal{L}\{\ldots\}$ differential operator
$\mathcal{L}$ matrix of differential operators

Abbreviations

1D   one dimensional
2D   two dimensional
BEM   boundary element method
c   centroid
EB   Euler-Bernoulli
e   elemental
el   elastic
f   flange
FDM   finite difference method
FEM   finite element method
FSDT   first-order shear deformation theory
FVM   finite volume method
geo   geometric
h   horizontal
L   Levinson
l   longitudinal
n   neutral
sc   shear center
SSDT   second-order shear deformation theory
T   Timoshenko
TSDT   third-order shear deformation theory
v   vertical
w   web

Chapter 1
Introduction to Continuum Mechanical Modeling

Abstract The first chapter introduces to major idea and the continuum mechanical background to model structural members. It is explained that physical problems are described based on differential equations. In the context of structural mechanics, these differential equations are obtained by combining the three basic equations of continuum mechanics, i.e., the kinematics relationship, the constitutive law, and the equilibrium equation. Furthermore, some explanations on the choice of the coordinate system for bending problems are provided.

Engineers describe physical phenomena and processes typically by equations, particularly by partial differential equations [2, 3, 16]. In this context, the derivation and the solution of these differential equations is the task of engineers, obviously requiring fundamental knowledge from physics and engineering mathematics.

The importance of partial differential equations is clearly represented in the following quote: 'For more than 250 years partial differential equations have been clearly the most important tool available to mankind in order to understand a large variety of phenomena, natural at first and then those originating from human activity and technological development. Mechanics, physics and their engineering applications were the first to benefit from the impact of partial differential equations on modeling and design, . . .' [5].

In the one-dimensional case, a physical problem can be generally described in a spatial domain Ω by the differential equation[1]

$$\mathcal{L}\{y(x)\} = b \quad (x \in \Omega) \tag{1.1}$$

and by the conditions which are prescribed on the boundary Γ. The differential equation is also called the *strong form* or the *original statement* of the problem. The expression 'strong form' comes from the fact that the differential equation describes exactly each point x in the domain of the problem. The operator $\mathcal{L}\{\dots\}$ in Eq. (1.1) is an arbitrary differential operator which can take, for example, the following forms:

---

[1]For some problems, Eq. (1.1) must be generalized to a system of differential equations.

$$\mathcal{L}\{\ldots\} = \frac{\mathrm{d}^2}{\mathrm{d}x^2}\{\ldots\}, \tag{1.2}$$

$$\mathcal{L}\{\ldots\} = \frac{\mathrm{d}^4}{\mathrm{d}x^4}\{\ldots\}, \tag{1.3}$$

$$\mathcal{L}\{\ldots\} = \frac{\mathrm{d}^4}{\mathrm{d}x^4}\{\ldots\} + \frac{\mathrm{d}}{\mathrm{d}x}\{\ldots\} + \{\ldots\}. \tag{1.4}$$

Furthermore, variable b in Eq. (1.1) is a given function, and in the case of $b = 0$, the equation reduces to the *homogeneous differential equation*: $\mathcal{L}\{y(x)\} = 0$. More specific expressions of Eqs. (1.3) till (1.4) can take the following forms [10]:

$$a\frac{\mathrm{d}^2 y(x)}{\mathrm{d}x^2} = b, \tag{1.5}$$

$$a\frac{\mathrm{d}^4 y(x)}{\mathrm{d}x^4} = b, \tag{1.6}$$

and will be used to describe the behavior of thin beams in the following sections.

Let us highlight at the end of this section that the derivations in the following chapters follow a common approach, see Fig. 1.1.

A combination of the kinematics equation (i.e., the relation between the strains and deformations) with the constitutive equation (i.e., the relation between the stresses and strains) and the equilibrium equation (i.e., the equilibrium between the internal reactions and the external loads) results in a partial differential equation. Limited to simple cases, analytical solutions are provided in this textbook. For more complex problems, numerical methods such as the finite difference method (FDM) [13], the finite element method (FEM) [12], the finite volume method (FVM) [14], or the boundary element method (BEM) [4] must be applied to derive approximate solutions of the partial differential equations.

The content of this monograph focuses on three different beam theories and applies the general scheme shown in Fig. 1.1 to derive the describing differential equations. The first theory is the theory for thin or slender beams according to Euler–Bernoulli. It is also called the shear-rigid beam theory. The historical development of this theory is described in [17, 20]. The following two theories relate to thick or compact beams, so-called shear-flexible beams, and go back to Timoshenko [18, 19] and Levinson [9]. Furthermore, it needs to be noted that the one-dimensional beam theories have corresponding counterparts in the two-dimensional space. In plate theories, the Euler-Bernoulli beam corresponds to the shear-rigid Kirchhoff plate and the Timoshenko beam corresponds to the shear-flexible Reissner–Mindlin plate, [6, 21].

To describe a bending problem in a single plane, there are different options on how to choose the coordinate system and the orientation of the single axes, see Fig. 1.2.

Let us have the beam in a horizontal position and the deflection should occur in the vertical direction. A choice common to mathematical education in secondary school would be to have the x-axis aligned with the horizontal longitudinal axis

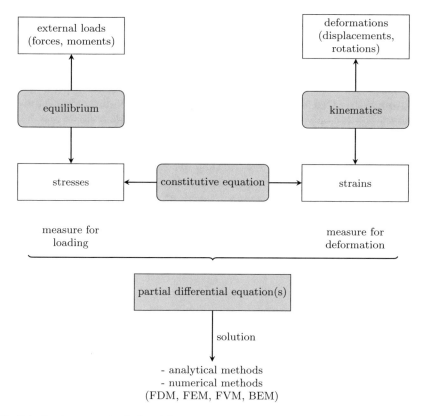

Fig. 1.1   Continuum mechanical modeling

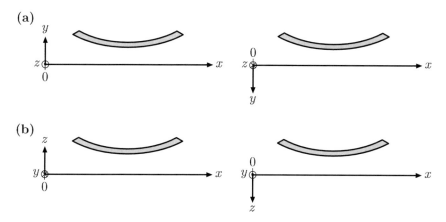

Fig. 1.2   Different ways to introduce beam bending problems: **a** x-y plane and **b** x-z plane

of the beam[2] and and the y-axis vertically upwards, see Fig. 1.2a. Civil engineers would rather prefer to have the y-axis vertically downwards, see Fig. 1.2b. This would give a positive deflection (downwards) under the influence of gravity. Alternatively to the configuration shown in Fig. 1.2a one may choose a x-z plane, see Fig. 1.2c. This choice has advantages if similarities between a bending beam and a bending plate should be shown since the thickness coordinate for two-dimensional structural members is commonly the z-coordinate. Obviously, the z-coordinate may be oriented downwards, see Fig. 1.2d. Nevertheless, an engineer must be able to master both configurations, i.e., bending in the x-y plane and the x-z plane since this relates to the general case of unsymmetrical bending, see Sects. 2.5, 3.5, and 4.5. The above mentioned options on how to introduce the coordinate system for single-plane bending problems is also reflected in common teaching books where bending is eater first introduced in the x-y plane (see for example [1, 8, 15]) or the x-z plane (see for example [7]).

References

1. Beer FP, Johnston ER Jr, DeWolf JT, Mazurek DF (2009) Mechanics of materials. McGraw-Hill, New York
2. Debnath L (2012) Nonlinear partial differential equations for scientists and engineers. Springer, New York
3. Formaggia L, Saleri F, Veneziani A (2012) Solving numerical PDEs: problems, applications, exercises. Springer, Milan
4. Gaul L, Kögl M, Wagner M (2003) Boundary element methods for engineers and scientists: an introductory course with advanced topics. Springer, Berlin
5. Glowinski R, Neittaanmäki P (eds) (2008) Partial differential equations: modelling and numerical simulation. Springer, Dordrecht
6. Gould PL (1988) Analysis of shells and plates. Springer, New York
7. Gross D, Hauger W, Schröder J, Wall WA, Bonet J (2011) Engineering mechanics 2: mechanics of materials. Springer, Berlin
8. Hibbeler RC (2011) Mechanics of materials. Prentice Hall, Singapore
9. Levinson M (1981) A new rectangular beam theory. J Sound Vib 74:81–87
10. Öchsner A (2014) Elasto-plasticity of frame structure elements: modeling and simulation of rods and beams. Springer, Berlin
11. Öchsner A, Öchsner M (2018) A first introduction to the finite element analysis program MSC Marc/Mentat. Springer, Cham
12. Öchsner A (2020) Computational statics and dynamics: an introduction based on the finite element method. Springer, Singapore
13. Öchsner A (2021) Structural mechanics with a pen: a guide to solve finite difference problems. Springer, Cham
14. Petrova R (2012) Finite volume method – powerful means of engineering design. InTech, Rijeka
15. Popov L (1990) Engineering mechanics of solids. Prentice-Hall, Englewood Cliffs
16. Salsa S (2008) Partial differential equations in action: from modelling to theory. Springer, Milano

---

[2]The common approach in analytical mechanics is to align the x-axis with the longitudinal axis of the beam. However, in the context of the finite element method, the local z-axis might be oriented along the element, see [11].

17. Szabó I (1996) Geschichte der mechaninschen Prinzipien. Birkhäuser Verlag, Basel
18. Timoshenko SP (1921) On the correction for shear of the differential equation for transverse vibrations of prismatic bars. Philos Mag 41:744–746
19. Timoshenko SP (1922) On the transverse vibrations of bars of uniform cross-section. Philos Mag 43:125–131
20. Timoshenko SP (1953) History of strength of materials. McGraw-Hill Book Company, New York
21. Timoshenko S, Woinowsky-Krieger S (1959) Theory of plates and shells. McGraw-Hill Book Company, New York

Chapter 2
Euler–Bernoulli Beam Theory

Abstract This chapter presents the analytical description of thin, or so-called shear-rigid, beam members according to the Euler–Bernoulli theory. Based on the three basic equations of continuum mechanics, i.e., the kinematics relationship, the constitutive law, and the equilibrium equation, the partial differential equations, which describe the physical problem, are presented. All equations are introduced for single plane bending in the x-y plane as well as the x-z plane. Analytical solutions of the partial differential equation are given for simple cases. In addition, this chapter treats the case of unsymmetrical bending, a superposition of tensile and bending modes as well as the shear stress distribution for solid section and thin-walled beams.

2.1 Introduction

A thin or Euler–Bernoulli beam is defined as a long prismatic body. The following derivations are restricted to some simplifications:

- only applying to straight beams,
- no elongation along the longitudinal (x) axis,
- no torsion around the longitudinal (x) axis,
- deformations in a single plane, i.e. symmetrical bending,
- small deformations, and
- simple cross sections.

The classic theories of beam bending distinguish between shear-rigid and shear-flexible models. The shear rigid-beam, also called the thin or Euler–Bernoulli beam,[1] neglects the shear deformation from the shear forces. This theory implies that a cross-sectional plane which was perpendicular to the beam axis before the deformation remains in the deformed state perpendicular to the beam axis, see Fig. 2.1a. Furthermore, it is assumed that a cross-sectional plane stays plane and unwarped in

---

[1] A historical analysis of the development of the classical beam theory and the contribution of different scientists can be found in [7].

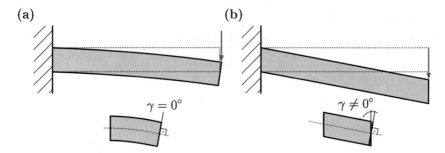

Fig. 2.1 Different deformation modes of a bending beam: **a** shear-rigid; **b** shear-flexible. Adapted from [6]

the deformed state. These two assumptions are also known as Bernoulli's hypothesis. Altogether one imagines that cross-sectional planes are rigidly fixed to the center line of the beam[2] so that a change of the center line affects the entire deformation. Consequently, it is also assumed that the geometric dimensions[3] of the cross-sectional planes do not change.

In the case of shear-flexible beams, for example the Timoshenko beam (see Chap. 3) or the Levinson beam (see Chap. 4), the shear deformation is considered in addition to the bending deformation and cross-sectional planes are rotated by an angle γ compared to the perpendicular line, see Fig. 2.1b. For homogeneous beams for which the length is 10 to 20 times larger than a characteristic dimension of the cross section, the shear fraction is usually disregarded in the first approximation. The different load types, meaning pure bending moment loading or shear due to shear force, lead to different stress fractions in a beam. In the case of an Euler–Bernoulli beam, deformation occurs solely through normal forces, which are linearly distributed over the cross section. Consequently, a tension—alternatively a compression maximum on the bottom face—maximum on the top face occurs, see Fig. 2.2a. In the case of symmetric cross sections, the zero crossing[4] occurs in the middle of the cross section. The shear stress distribution for a rectangular cross section is parabolic at which the maximum occurs at the neutral axis and is zero at both the top and bottom surface, see Fig. 2.2b. This shear stress distribution can be calculated for the Euler–Bernoulli beam but is not considered for the derivation of the deformation (i.e., the bending line) or the evaluation of stress failure criteria.

Finally, it needs to be noted that the one-dimensional beam theories have corresponding counterparts in the two-dimensional space, see Table 2.1. In plate theories, the Euler–Bernoulli beam corresponds to the shear-rigid Kirchhoff plate and the Timoshenko beam corresponds to the shear-flexible Reissner–Mindlin plate, [1, 4, 14].

---

[2]More precisely, this is the neutral fibre or the bending line.

[3]Consequently, the width b and the height h of a, for example, rectangular cross section remain the same.

[4]The sum of all points with $\sigma = 0$ along the beam axis is called the neutral fiber..

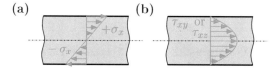

Fig. 2.2 Different stress distributions of a beam with rectangular cross section and linear-elastic material behavior: **a** normal stress and **b** shear stress

Table 2.1 Analogies between the beam and plate theories

	Beam theory	Plate theory
Dimensionality	1D	2D
Shear-rigid	Euler–Bernoulli beam	Kirchhoff plate
Shear-flexible	Timoshenko beam	Reissner–Mindlin plate
	Levinson beam	see [12]

Further details regarding the beam theory and the corresponding basic definitions and assumptions can be found in references [3, 5, 8, 13]. In the following sections, only the Euler–Bernoulli beam is considered. Consideration of the shear part takes place in Chaps. 3 and 4.

2.2 Deformation in the x-y Plane

The external loads, which are considered within this section, are single forces F_y, single moments M_z, distributed loads $q_y(x)$, and distributed moments $m_z(x)$, see Fig. 2.3. These loads have in common that their line of action (force) or the direction of the momentum vector are orthogonal to the longitudinal (x) axis of the beam and cause its bending.

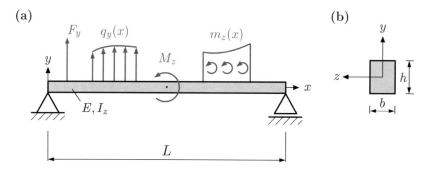

Fig. 2.3 General configuration for thin beam problems in the x-y plane: **a** example of boundary conditions and external loads (drawn in their positive directions); **b** cross-sectional area

2.2.1 Kinematics

For the derivation of the kinematics relation, a beam with length L is under constant moment loading $M_z(x) = $ const., meaning under *pure* bending, is considered, see Fig. 2.4. One can see that both external single moments at the left- and right-hand boundary lead to a positive bending moment distribution M_z within the beam. The vertical position of a point with respect to the center line of the beam *without action* of an external load is described through the y-coordinate. The vertical *displacement* of a point on the center line of the beam, meaning for a point with $y = 0$, under action of the external load is indicated with u_y. The deformed center line is represented by the sum of these points with $y = 0$ and is referred to as the bending line $u_y(x)$.

In the case of a deformation in the x-y plane, it is important to check the positive orientation of the internal reactions, the positive rotational angle, and the slope see Fig. 2.5. The internal reactions at the right-hand boundary are directed in the positive directions of the coordinate axes. Thus, a positive moment at the right-hand boundary is counterclockwise oriented (as the positive rotational angle), see Fig. 2.5a. Furthermore, the slope is positive, see Fig. 2.5b. This must be considered during the derivations of the corresponding equations.

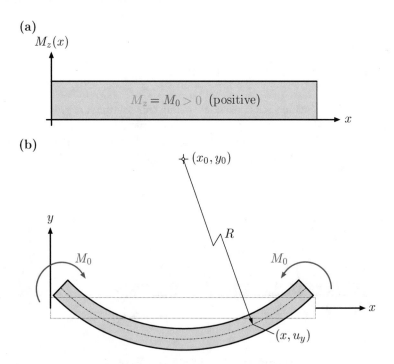

Fig. 2.4 Beam under pure bending in the x-y plane: **a** moment distribution; **b** deformed beam. Note that the deformation is exaggerated for better illustration. For the deformations considered in this chapter the following applies: $R \gg L$

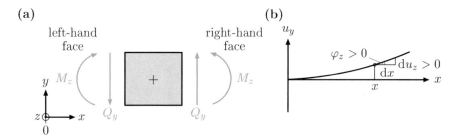

Fig. 2.5 Positive definition of **a** internal reactions and **b** rotation (positive slope) in the x-y plane

Only the center line of the deformed beam is considered in the following. Through the relation for an arbitrary point (x, u_y) on a circle with radius R around the center point (x_0, y_0), meaning

$$(x - x_0)^2 + (u_y(x) - y_0)^2 = R^2 , \tag{2.1}$$

one obtains through differentiation with respect to the x-coordinate

$$2(x - x_0) + 2(u_y(x) - y_0)\frac{du_y(x)}{dx} = 0 , \tag{2.2}$$

alternatively after another differentiation:

$$2 + 2\frac{du_y}{dx}\frac{du_y}{dx} + 2(u_y(x) - y_0)\frac{d^2u_y}{dx^2} = 0 . \tag{2.3}$$

Equation (2.3) provides the vertical distance between an arbitrary point on the center line of the beam and the center point of a circle as

$$(u_y - y_0) = -\frac{1 + \left(\dfrac{du_y}{dx}\right)^2}{\dfrac{d^2u_y}{dx^2}} , \tag{2.4}$$

while the difference between the x-coordinates results from Eq. (2.2):

$$(x - x_0) = -(u_y - y_0)\frac{du_y}{dx} . \tag{2.5}$$

If the expression according to Eq. (2.4) is used in Eq. (2.5) the following results:

$$(x - x_0) = \frac{\mathrm{d}u_y}{\mathrm{d}x} \frac{1 + \left(\dfrac{\mathrm{d}u_y}{\mathrm{d}x}\right)^2}{\dfrac{\mathrm{d}^2 u_y}{\mathrm{d}x^2}}. \tag{2.6}$$

Inserting both expressions for the x- and y-coordinate differences according to Eqs. (2.6) and (2.4) in the circle equation according to (2.1) leads to:

$$R^2 = (x - x_0)^2 + (u_y - y_0)^2$$

$$= \left(\frac{\mathrm{d}u_y}{\mathrm{d}x}\right)^2 \frac{\left(1 + \left(\frac{\mathrm{d}u_y}{\mathrm{d}x}\right)^2\right)^2}{\left(\frac{\mathrm{d}^2 u_y}{\mathrm{d}x^2}\right)^2} + \frac{\left(1 + \left(\frac{\mathrm{d}u_y}{\mathrm{d}x}\right)^2\right)^2}{\left(\frac{\mathrm{d}^2 u_y}{\mathrm{d}x^2}\right)^2} \tag{2.7}$$

$$= \left(\left(\frac{\mathrm{d}^2 u_y}{\mathrm{d}x^2}\right)^2 + 1\right) \frac{\left(1 + \left(\frac{\mathrm{d}u_y}{\mathrm{d}x}\right)^2\right)^2}{\left(\frac{\mathrm{d}^2 u_y}{\mathrm{d}x^2}\right)^2}$$

$$= \frac{\left(1 + \left(\frac{\mathrm{d}u_y}{\mathrm{d}x}\right)^2\right)^3}{\left(\frac{\mathrm{d}^2 u_y}{\mathrm{d}x^2}\right)^2}. \tag{2.8}$$

Thus, the radius of curvature is obtained as:

$$|R| = \frac{\left(1 + \left(\frac{\mathrm{d}u_y}{\mathrm{d}x}\right)^2\right)^{3/2}}{\left|\frac{\mathrm{d}^2 u_y}{\mathrm{d}x^2}\right|}. \tag{2.9}$$

To decide if the radius of curvature is positive or negative, let us have a look at Fig. 2.6 where a curve with its tangential and normal vectors is shown. Since the curve in the configuration of Fig. 2.6b is bending toward to the normal vector $\boldsymbol{n}$, it holds that $\frac{\mathrm{d}^2 u_y}{\mathrm{d}x^2} > 0$ and the radius of curvature is obtained for a positive bending moment as:

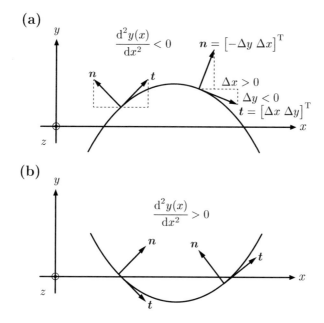

Fig. 2.6 Sign of the second-order derivative of a curve in the *x*-*y* plane: **a** a curve which is bending away from normal vector **n** gives a negative sign and **b** a curve which is bending towards the normal vector **n** gives a positive sign

$$R = +\frac{\left(1 + \left(\dfrac{\mathrm{d}u_y}{\mathrm{d}x}\right)^2\right)^{3/2}}{\dfrac{\mathrm{d}^2 u_y}{\mathrm{d}x^2}}. \tag{2.10}$$

Note that the expression curvature, which results as a reciprocal value from the curvature radius, $\kappa = \frac{1}{R}$, is used as well.

For small bending deflections, meaning $u_y \ll L$, $\frac{\mathrm{d}u_y}{\mathrm{d}x} \ll 1$ results and Eq. (2.10) simplifies to:

$$R = +\frac{1}{\dfrac{\mathrm{d}^2 u_y}{\mathrm{d}x^2}} \quad \text{or} \quad \kappa = \frac{1}{R} = +\frac{\mathrm{d}^2 u_y}{\mathrm{d}x^2}. \tag{2.11}$$

For the determination of the normal strain, one refers to its general definition, meaning elongation referring to initial length. Relating to the configuration shown in Fig. 2.7, the longitudinal elongation of a fibre at distance *y* to the neutral fibre allows to express the strain as:

$$\varepsilon_x = \frac{\mathrm{d}s - \mathrm{d}x}{\mathrm{d}x}. \tag{2.12}$$

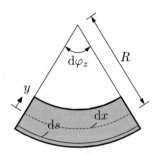

The lengths of the circular arcs ds and dx result from the corresponding radii and
the enclosed angles in radian measure as:

$$dx = Rd\varphi_z, \tag{2.13}$$

$$ds = (R - y)d\varphi_z. \tag{2.14}$$

If these relations for the circular arcs are used in Eq. (2.12), the following results:

$$\varepsilon_x = \frac{(R - y)d\varphi_z - Rd\varphi_z}{dx} = -y\frac{d\varphi_z}{dx}. \tag{2.15}$$

It results from Eq. (2.13) that $\frac{d\varphi_z}{dx} = \frac{1}{R}$ and together with relation (2.51) the strain
can finally be expressed as follows:

$$\varepsilon_x(x, y) = -y\frac{1}{R} \overset{(2.11)}{=} -y\frac{d^2u_y(x)}{dx^2} \overset{(2.11)}{=} -y\kappa. \tag{2.16}$$

An alternative derivation of the kinematics relation results from consideration of
Fig. 2.8. From the relation of the right-angled triangle $0'1'2'$, this means[5] $\sin\varphi_z = \frac{-u_x}{y}$, the following relation results for small angles ($\sin\varphi_z \approx \varphi_z$):

$$u_x = -y\varphi_z. \tag{2.17}$$

Furthermore, it holds that the rotation angle of the slope equals the center line for
small angles:

$$\tan\varphi_z = \frac{+du_y(x)}{dx} \approx \varphi_z. \tag{2.18}$$

If Eqs. (2.18) and (2.17) are combined, the following results:

---

[5]Note that according to the assumptions of the Euler–Bernoulli beam the lengths 01 and $0'1'$ remain
unchanged.

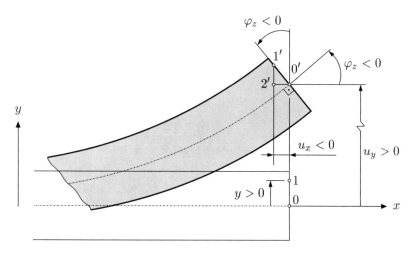

Fig. 2.8 Alternative configuration for the derivation of the kinematics relation for bending in the x-y plane. Note that the deformation is exaggerated for better illustration

$$u_x = -y \frac{du_y(x)}{dx}.$$ (2.19)

The last relation equals $(ds - dx)$ in Eq. (2.12) and differentiation with respect to the x-coordinate leads directly to Eq. (2.16).

2.2.2 Constitutive Equation

The one-dimensional Hooke's law can be assumed in the case of the bending beam, since, according to the requirement, only normal stresses are regarded in this section:

$$\sigma_x = E \varepsilon_x,$$ (2.20)

where E is Young's modulus. Through the kinematics relation according to Eq. (2.16), the stress results as a function of the deflection to:

$$\sigma_x(x, y) = -E y \frac{d^2 u_y(x)}{dx^2}.$$ (2.21)

The stress distribution shown in Fig. 2.9a generates the internal moment, which acts in this cross section. To calculate this internal moment, the stress is multiplied by a surface element, so that the resulting force is obtained. Multiplication with the corresponding lever arm then gives the internal moment. Since the stress is linearly

(a)   (b)

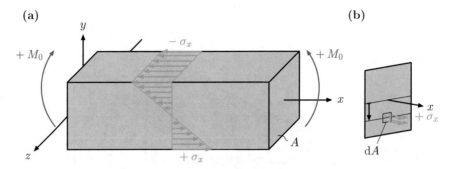

Fig. 2.9 a Schematic representation of the normal stress distribution $\sigma_x = \sigma_x(y)$ of a bending beam; **b** Definition and position of an infinitesimal surface element for the derivation of the resulting moment action due to the normal stress distribution

distributed over the height, the evaluation is done for an infinitesimally small surface element:

$$\mathrm{d}M_z = (-y)(+\sigma_x)\mathrm{d}A = -y\sigma_x\mathrm{d}A .\tag{2.22}$$

Therefore, the entire moment results via integration over the entire surface in:

$$M_z = -\int_A y\sigma_x\mathrm{d}A \stackrel{(2.21)}{=} +\int_A yEy\frac{\mathrm{d}^2u_y(x)}{\mathrm{d}x^2}\mathrm{d}A .\tag{2.23}$$

Assuming that the Young's modulus is constant, the internal moment around the z-axis results in:

$$M_z = E\frac{\mathrm{d}^2u_y}{\mathrm{d}x^2}\underbrace{\int_A y^2\mathrm{d}A}_{I_z} = \frac{I_z\sigma_x}{z} .\tag{2.24}$$

The integral in Eq. (2.24) is the so-called   axial second moment of area or axial surface moment of 2nd order in the SI unit m^4. This factor is only dependent on the geometry of the cross section and is also a measure of the stiffness of a plane cross section against bending. The values of the axial second moment of area for simple geometric cross sections are collected in Table 2.2.

Consequently the internal moment can also be expressed as

$$M_z = EI_z\frac{\mathrm{d}^2u_y}{\mathrm{d}x^2} \stackrel{(2.11)}{=} \frac{EI_z}{R} = EI_z\kappa.\tag{2.25}$$

Equation (2.25) describes the bending line $u_y(x)$ as a function of the bending moment and is therefore also referred to as the bending line-moment relation. The product EI_z in Eq. (2.25) is also called the bending stiffness. If the result from Eq. (2.25) is

Table 2.2 Axial second moment of area around the *z*-axis

Cross section	I_z
(circle, diameter $D = 2R$, axes z and y)	$\dfrac{\pi D^4}{64} = \dfrac{\pi R^4}{4}$
(ellipse, height $2a$, width $2b$)	$\dfrac{\pi ab^3}{4}$
(square, side $a \times a$)	$\dfrac{a^4}{12}$
(rectangle, height h, width b)	$\dfrac{hb^3}{12}$
(right triangle, base b, height h, $b/3$, $h/3$)	$\dfrac{hb^3}{36}$
(isosceles triangle, base b, height h, $h/3$)	$\dfrac{bh^3}{48}$

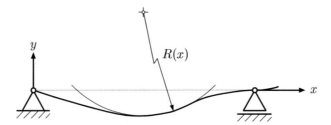

Fig. 2.10 Deformation of a beam in the x-y plane with $M_z(x) \neq$ const.

used in the relation for the bending stress according to Eq. (2.21), the distribution of stress over the cross section results in:

$$\sigma_x(x, y) = -\frac{M_z(x)}{I_z} y(x) \,. \tag{2.26}$$

The minus sign in Eq. (2.26) causes that a positive bending moment (see Fig. 2.4) leads to a tensile stress in the lower beam half (meaning for $y < 0$).

In the case of plane bending with $M_z(x) \neq$ const., the bending line can be approximated in each case locally through a circle of curvature, see Fig. 2.10. Therefore, the result for *pure* bending according to Eq. (2.25) can be transferred to the case of plane bending as:

$$EI_z \frac{d^2 u_y(x)}{dx^2} = M_z(x) \,. \tag{2.27}$$

Let us note at the end of this section that Hooke's law in the form of Eq. (2.20) is not so easy to apply[6] in the case of beams since the stress and strain is linearly changing over the height of the cross section, see Eq. (2.26) and Fig. 2.9. Thus, it might be easier to apply a so-called stress resultant or generalized stress, i.e. a simplified representation of the normal stress state[7] based on the acting bending moment:

$$M_z(x) = -\iint y\sigma_x(x, y) \, dA \,, \tag{2.28}$$

which was already introduced in Eq. (2.22). Using in addition the curvature[8] $\kappa = \kappa(x)$ (see Eq. (2.16)) instead of the strain $\varepsilon_x = \varepsilon_x(x, y)$, the constitutive equation can be easier expressed as shown in Fig. 2.11. The variables M_z and κ have both the advantage that they are constant for any location x of the beam.

---

[6]However, this formulation works well in the case of rod elements since stress and strain are constant over the cross section, i.e. $\sigma_x = \sigma_x(x)$ and $\varepsilon_x = \varepsilon_x(x)$.

[7]A similar stress resultant can be stated for the shear stress based on the shear force: $Q_y(x) = \iint \tau_{xy}(x, y) \, dA$.

[8]The curvature is then called a generalized strain.

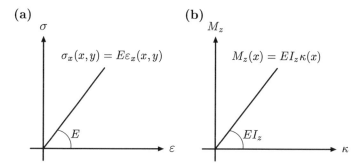

Fig. 2.11 Formulation of the constitutive law based on **a** stress and **b** stress resultant (bending in the x-y plane)

2.2.3 Equilibrium

The equilibrium conditions are derived from an infinitesimal beam element of length dx, which is loaded by a constant distributed force $q_{y,0}$ as well as a constant distributed moment $m_{z,0}$, see Fig. 2.12. The internal reactions are drawn on both cut faces, i.e. at locations x and $x + dx$. One can see that a positive shear force is oriented in the positive y-direction at the right-hand face[9] and that a positive bending moment has the same rotational direction as the positive z-axis (right-hand grip rule[10]). The orientation of shear force and bending moment is reversed at the left-hand, i.e. negative, face in order to cancel in sum the effect of the internal reactions at both faces. This convention for the direction of the internal directions is maintained in the following. Furthermore, it can be derived from Fig. 2.12 that an upwards directed *external* force or alternatively a mathematically positive oriented *external* moment at the right-hand face leads to a positive internal shear force or alternatively a positive internal bending moment. In a corresponding way, it results that a downwards directed *external* force or alternatively a mathematically negative oriented *external* moment at the left-hand face leads to a positive internal shear force or alternatively a positive internal bending moment.

The equilibrium condition will be determined in the following for the vertical forces. Assuming that forces in the direction of the positive y-axis are considered positive, the following results:

$$- Q_y(x) + Q_y(x + dx) + q_{y,0}dx = 0 \,. \tag{2.29}$$

---

[9] A positive cut face is defined by the surface normal on the cut plane which has the same orientation as the positive x-axis. It should be regarded that the surface normal is always directed outward.

[10] If the axis is grasped with the right hand in a way so that the spread out thumb points in the direction of the positive axis, the bent fingers then show the direction of the positive rotational direction.

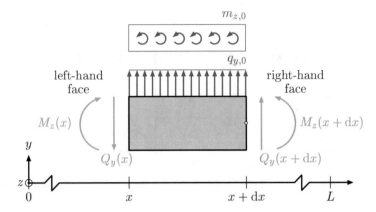

Fig. 2.12 Infinitesimal beam element in the x-y plane with internal reactions and constant distributed loads

If the shear force on the right-hand face is expanded in a Taylor's series of first order, meaning

$$Q_y(x + \mathrm{d}x) \approx Q_y(x) + \frac{\mathrm{d}Q_y(x)}{\mathrm{d}x}\mathrm{d}x , \qquad (2.30)$$

Equation (2.29) results in

$$- Q_y(x) + Q_y(x) + \frac{\mathrm{d}Q_y(x)}{\mathrm{d}x}\mathrm{d}x + q_{y,0}\mathrm{d}x = 0 , \qquad (2.31)$$

or alternatively after simplification finally to:

$$\frac{\mathrm{d}Q_y(x)}{\mathrm{d}x} = -q_{y,0} . \qquad (2.32)$$

For the special case that no distributed load is acting ($q_{y,0} = 0$), Eq. (2.32) simplifies to:

$$\frac{\mathrm{d}Q_y(x)}{\mathrm{d}x} = 0 . \qquad (2.33)$$

The equilibrium of moments around the reference point at $x + \mathrm{d}x$ gives:

$$- M_z(x + \mathrm{d}x) + M_z(x) - Q_y(x)\mathrm{d}x + \frac{1}{2}q_{y,0}\mathrm{d}x^2 - m_{z,0}\mathrm{d}x = 0 . \qquad (2.34)$$

If the bending moment on the right-hand face is expanded into a Taylor's series of first order similar to Eq. (2.30) and consideration that the term $\frac{1}{2}q_{y,0}\mathrm{d}x^2$ as infinitesimal small size of higher order can be disregarded, finally the following results:

$$\frac{\mathrm{d}M_z(x)}{\mathrm{d}x} = -Q_y(x) - m_{z,0} . \qquad (2.35)$$

Table 2.3 Elementary basic equations for the bending of a thin beam in the x-y plane. The differential equations are given under the assumption of constant bending stiffness EI_z

Name	Equation
Kinematics	$\varepsilon_x(x, y) = -y\dfrac{\mathrm{d}^2 u_y(x)}{\mathrm{d}x^2}$
Equilibrium	$\dfrac{\mathrm{d}Q_y(x)}{\mathrm{d}x} = -q_y(x);$ $\dfrac{\mathrm{d}M_z(x)}{\mathrm{d}x} = -Q_y(x) - m_z(x)$
Constitutive equation	$\sigma_x(x, y) = E\varepsilon_x(x, y)$
Stress	$\sigma_x(x, y) = -\dfrac{M_z(x)}{I_z}y(x)$
Diff'equation	$EI_z\dfrac{\mathrm{d}^2 u_y(x)}{\mathrm{d}x^2} = M_z(x)$
	$EI_z\dfrac{\mathrm{d}^3 u_y(x)}{\mathrm{d}x^3} = -Q_y(x) - m_z(x)$
	$EI_z\dfrac{\mathrm{d}^4 u_y(x)}{\mathrm{d}x^4} = q_y(x) - \dfrac{\mathrm{d}m_z(x)}{\mathrm{d}x}$

The combination of Eqs. (2.32) and (2.35) leads to the relation between the bending moment and the distributed loads:

$$\frac{\mathrm{d}^2 M_z(x)}{\mathrm{d}x^2} = -\frac{\mathrm{d}Q_y(x)}{\mathrm{d}x} - \frac{\mathrm{d}m_{z,0}}{\mathrm{d}x} = q_{y,0} - \frac{\mathrm{d}m_{z,0}}{\mathrm{d}x}. \tag{2.36}$$

Finally, the elementary basic equations for the bending of a beam in the x-y plane for arbitrary moment loading $M_z(x)$ are summarized in Table 2.3.

2.2.4 Differential Equation

Different formulations of the fourth-order differential equation are collected in Table 2.4 where different types of loadings, geometry and bedding are differentiated. The last case in Table 2.4 refers to the elastic foundation of a beam which is also know in the literature as Winkler foundation [15]. The elastic foundation or Winkler foundation modulus k has in the case of beams[11] the unit of force per unit area.

---

[11] In the general case, the unit of the elastic foundation modulus is force per unit area per unit length, i.e. $\frac{\mathrm{N}}{\mathrm{m}^2}/\mathrm{m} = \frac{\mathrm{N}}{\mathrm{m}^3}$.

Table 2.4 Different formulations of the partial differential equation for a thin beam in the x-y plane (x-axis: right facing; y-axis: upward facing)

Configuration	Partial differential equation
E, I_z	$EI_z \dfrac{\mathrm{d}^4 u_y(x)}{\mathrm{d}x^4} = 0$
$E(x), I_z(x)$	$\dfrac{\mathrm{d}^2}{\mathrm{d}x^2}\left(E(x)I_z(x)\dfrac{\mathrm{d}^2 u_y(x)}{\mathrm{d}x^2}\right) = 0$
$q_y(x)$	$EI_z \dfrac{\mathrm{d}^4 u_y(x)}{\mathrm{d}x^4} = q_y(x)$
$m_z(x)$	$EI_z \dfrac{\mathrm{d}^4 u_y(x)}{\mathrm{d}x^4} = -\dfrac{\mathrm{d}m_z(x)}{\mathrm{d}x}$
$k(x)$	$EI_z \dfrac{\mathrm{d}^4 u_y(x)}{\mathrm{d}x^4} = -k(x)u_y(x)$

If we replace the common formulations of the second- and first-order derivatives, i.e. $\frac{\mathrm{d}^2 \dots}{\mathrm{d}x^2}$ and $\frac{\mathrm{d}\dots}{\mathrm{d}x}$, by formal operator symbols, i.e. $\mathcal{L}_2(\dots)$ and $\mathcal{L}_1(\dots)$, the basic equations can be stated in a more formal way as given in Table 2.5. Such a general formulation is suitable for the derivation of the principal finite element equation based on the weighted residual method [10, 11].

Under the assumption of constant material ($E = $ const.) and geometric ($I_z = $ const.) properties, the differential equation in Table 2.5 can be integrated four times for constant distributed loads ($q_y(x) = q_0 = $ const., $m_z(x) = 0$) to obtain the general analytical solution of the problem [9]:

$$u_y(x) = \frac{1}{EI_z}\left(\frac{q_0 x^4}{24} + \frac{c_1 x^3}{6} + \frac{c_2 x^2}{2} + c_3 x + c_4 \right), \qquad (2.37)$$

where the four constants of integration c_i ($i = 1, \dots, 4$) must be determined based on the boundary conditions. The following equations for the shear force $Q_y(x)$, the bending moment $M_z(x)$, and the rotation $\varphi_z(x)$ were obtained based on one-, two- and three-times integration and might be useful to determine some of the constants of integration:

Table 2.5 Different formulations of the basic equations for a thin beam (bending in the x-y plane; x-axis along the principal beam axis). E: Young's modulus; I_z: second moment of area; q_y: length-specific distributed force; m_z: length-specific distributed moment; $\mathcal{L}_2 = \frac{\mathrm{d}^2(...)}{\mathrm{d}x^2}$: second-order derivative; $\mathcal{L}_1 = \frac{\mathrm{d}(...)}{\mathrm{d}x}$: first-order derivative

Specific formulation	General formulation [2]
Kinematics	
$\varepsilon_x(x, y) = -y\dfrac{\mathrm{d}^2 u_y(x)}{\mathrm{d}x^2}$	$\varepsilon_x(x, y) = -y\mathcal{L}_2\left(u_y(x)\right)$
$\kappa(x) = -\dfrac{\mathrm{d}^2 u_y(x)}{\mathrm{d}x^2}$	$\kappa(x) = -\mathcal{L}_2\left(u_y(x)\right)$
Constitution	
$\sigma_x(x, y) = E\varepsilon_x(x, y)$	$\sigma_x(x, y) = C\varepsilon_x(x, y)$
$M_z(x) = EI_z\kappa(x)$	$M_z(x) = D\kappa(x)$
Equilibrium	
$\dfrac{\mathrm{d}^2 M_z(x)}{\mathrm{d}x^2} - q_y(x) + \dfrac{\mathrm{d}m_z(x)}{\mathrm{d}x} = 0$	$\mathcal{L}_2^{\mathrm{T}}(M_z(x)) - q_y(x) + \mathcal{L}_1^{\mathrm{t}}m_z(x)) = 0$
PDE	
$\dfrac{\mathrm{d}^2}{\mathrm{d}x^2}\left(EI_z\dfrac{\mathrm{d}^2 u_y(x)}{\mathrm{d}x^2}\right) - q_y(x) + \dfrac{\mathrm{d}m_z(x)}{\mathrm{d}x} = 0$	$\mathcal{L}_2^{\mathrm{T}}(D\mathcal{L}_2(u_y(x))) - q_y(x) + \mathcal{L}_1^{\mathrm{t}}m_z(x)) = 0$

$$Q_y(x) = -q_0 x - c_1 \,, \tag{2.38}$$

$$M_z(x) = \frac{q_0 x^2}{2} + c_1 x + c_2 \,, \tag{2.39}$$

$$\varphi_z(x) = +\frac{\mathrm{d}u_y(x)}{\mathrm{d}x} = \frac{1}{EI_z}\left(\frac{q_0 x^3}{6} + \frac{c_1 x^2}{2} + c_2 x + c_3\right) . \tag{2.40}$$

2.3 Deformation in the x-z Plane

The external loads, which are considered within this section, are single forces F_z, single moments M_y, distributed loads $q_z(x)$, and distributed moments $m_y(x)$, see Fig. 2.13. These loads have in common that their line of action (force) or the direction of the momentum vector are orthogonal to the longitudinal (x) axis of the beam and cause its bending.

2.3.1 Kinematics

For the derivation of the kinematics relation, a beam with length L is under constant moment loading $M_y(x) = \text{const.}$, meaning under *pure* bending, is considered, see

(a)

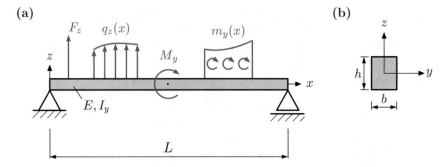

(b)

Fig. 2.13 General configuration for thin beam problems in the x-z plane: **a** example of boundary conditions and external loads (drawn in their positive directions); **b** cross-sectional area

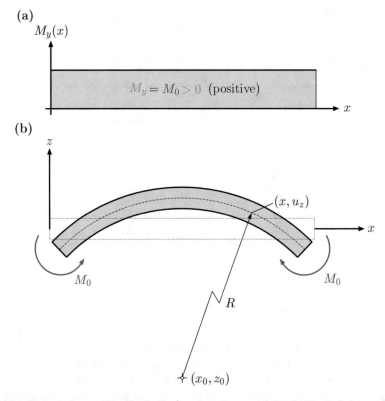

Fig. 2.14 Beam under pure bending in the x-z plane: **a** moment distribution; **b** deformed beam. Note that the deformation is exaggerated for better illustration. For the deformations considered in this chapter the following applies: $R \gg L$

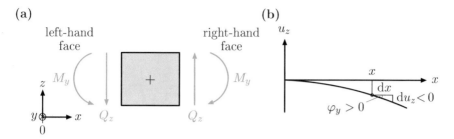

Fig. 2.15 Positive definition of **a** internal reactions and **b** rotation (but negative slope) in the x-z plane

Fig. 2.14. One can see that both external single moments at the left- and right-hand boundary lead to a positive bending moment distribution M_y within the beam. The vertical position of a point with respect to the center line of the beam *without action* of an external load is described through the z-coordinate. The vertical *displacement* of a point on the center line of the beam, meaning for a point with $z = 0$, under action of the external load is indicated with u_z. The deformed center line is represented by the sum of these points with $z = 0$ and is referred to as the bending line $u_z(x)$.

In the case of a deformation in the x-z plane, it is important to precisely distinguish between the positive orientation of the internal reactions, the positive rotational angle, and the slope see Fig. 2.15. The internal reactions at the right-hand boundary are directed in the positive directions of the coordinate axes. Thus, a positive moment at the right-hand boundary is clockwise oriented (as the positive rotational angle), see Fig. 2.15a. However, the slope is negative, see Fig. 2.15b. This difference requires some careful derivations of the corresponding equations.

Only the center line of the deformed beam is considered in the following. Through the relation for an arbitrary point (x, u_z) on a circle with radius R around the center point (x_0, z_0), meaning

$$(x - x_0)^2 + (u_z(x) - z_0)^2 = R^2 , \tag{2.41}$$

one obtains through differentiation with respect to the x-coordinate

$$2(x - x_0) + 2(u_z(x) - z_0)\frac{du_z(x)}{dx} = 0 , \tag{2.42}$$

alternatively after another differentiation:

$$2 + 2\frac{du_z}{dx}\frac{du_z}{dx} + 2(u_z(x) - z_0)\frac{d^2u_z}{dx^2} = 0 . \tag{2.43}$$

Equation (2.43) provides the vertical distance between an arbitrary point on the center line of the beam and the center point of a circle as

$$(u_z - z_0) = -\frac{1 + \left(\dfrac{du_z}{dx}\right)^2}{\dfrac{d^2 u_z}{dx^2}}, \tag{2.44}$$

while the difference between the x-coordinates results from Eq. (2.42):

$$(x - x_0) = -(u_z - z_0)\frac{du_z}{dx}. \tag{2.45}$$

If the expression according to Eq. (2.44) is used in Eq. (2.45) the following results:

$$(x - x_0) = \frac{du_z}{dx}\frac{1 + \left(\dfrac{du_z}{dx}\right)^2}{\dfrac{d^2 u_z}{dx^2}}. \tag{2.46}$$

Inserting both expressions for the x- and z-coordinate differences according to Eqs. (2.46) and (2.44) in the circle equation according to (2.41) leads to:

$$R^2 = (x - x_0)^2 + (u_z - z_0)^2$$

$$= \left(\frac{du_z}{dx}\right)^2 \frac{\left(1 + \left(\frac{du_z}{dx}\right)^2\right)^2}{\left(\frac{d^2 u_z}{dx^2}\right)^2} + \frac{\left(1 + \left(\frac{du_z}{dx}\right)^2\right)^2}{\left(\frac{d^2 u_z}{dx^2}\right)^2} \tag{2.47}$$

$$= \left(\left(\frac{d^2 u_z}{dx^2}\right)^2 + 1\right)\frac{\left(1 + \left(\frac{du_z}{dx}\right)^2\right)^2}{\left(\frac{d^2 u_z}{dx^2}\right)^2}$$

$$= \frac{\left(1 + \left(\frac{du_z}{dx}\right)^2\right)^3}{\left(\frac{d^2 u_z}{dx^2}\right)^2}. \tag{2.48}$$

Thus, the radius of curvature is obtained as:

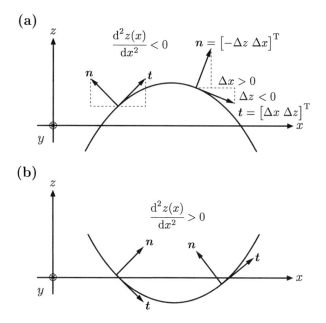

Fig. 2.16 Sign of the second-order derivative of a curve in the x-z plane: **a** a curve which is bending away from normal vector $\mathbf{n}$ gives a negative sign and **b** a curve which is bending towards the normal vector $\mathbf{n}$ gives a positive sign

$$|R| = \frac{\left(1 + \left(\dfrac{du_z}{dx}\right)^2\right)^{3/2}}{\left|\dfrac{d^2 u_z}{dx^2}\right|}. \tag{2.49}$$

To decide if the radius of curvature is positive or negative, let us have a look at Fig. 2.16 where a curve with its tangential and normal vectors is shown. Since the curve in the configuration of Fig. 2.16a is bending away from the normal vector $\mathbf{n}$, it holds that $\frac{d^2 u_z}{dx^2} < 0$ and the radius of curvature is obtained for a positive bending moment as:

$$R = -\frac{\left(1 + \left(\dfrac{du_z}{dx}\right)^2\right)^{3/2}}{\dfrac{d^2 u_z}{dx^2}}. \tag{2.50}$$

Note that the expression curvature, which results as a reciprocal value from the curvature radius, $\kappa = \frac{1}{R}$, is used as well.

Fig. 2.17 Segment of a
beam under pure bending in
the x-z plane. Note that the
deformation is exaggerated
for better illustration

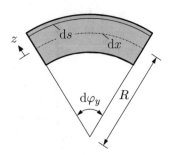

For small bending deflections, meaning $u_z \ll L$, $\frac{du_z}{dx} \ll 1$ results and Eq. (2.50)
simplifies to:

$$R = -\frac{1}{\dfrac{d^2 u_z}{dx^2}} \quad \text{or} \quad \kappa = \frac{1}{R} = -\frac{d^2 u_z}{dx^2}. \tag{2.51}$$

For the determination of the strain, one refers to its general definition, meaning
elongation referring to initial length. Relating to the configuration shown in Fig. 2.17,
the longitudinal elongation of a fibre at distance z to the neutral fibre allows to express
the strain as:

$$\varepsilon_x = \frac{ds - dx}{dx}. \tag{2.52}$$

The lengths of the circular arcs ds and dx result from the corresponding radii and
the enclosed angles in radian measure as:

$$dx = R d\varphi_y, \tag{2.53}$$

$$ds = (R + z) d\varphi_y. \tag{2.54}$$

If these relations for the circular arcs are used in Eq. (2.52), the following results:

$$\varepsilon_x = \frac{(R + z)d\varphi_y - R d\varphi_y}{dx} = z\frac{d\varphi_y}{dx}. \tag{2.55}$$

From Eq. (2.53) $\frac{d\varphi_y}{dx} = \frac{1}{R}$ results and together with relation (2.51) the strain can
finally be expressed as follows:

$$\varepsilon_x(x, z) = z\frac{1}{R} \overset{(2.51)}{=} -z\frac{d^2 u_z(x)}{dx^2} \overset{(2.51)}{=} z\kappa. \tag{2.56}$$

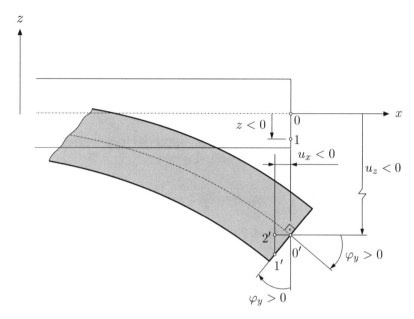

Fig. 2.18 Alternative configuration for the derivation of the kinematics relation for bending in the x-z plane. Note that the deformation is exaggerated for better illustration

An alternative derivation of the kinematics relation results from consideration of Fig. 2.18. From the relation of the right-angled triangle $0'1'2'$, this means[12] $\sin \varphi_y = \frac{u_x}{z}$, the following relation results for small angles ($\sin \varphi_y \approx \varphi_y$):

$$u_x = +z\varphi_y . \tag{2.57}$$

Furthermore, it holds that the rotation angle of the slope equals the center line for small angles:

$$\tan \varphi_y = \frac{-\,\mathrm{d}u_z(x)}{\mathrm{d}x} \approx \varphi_y . \tag{2.58}$$

If Eqs. (2.58) and (2.57) are combined, the following results:

$$u_x = -z\frac{\mathrm{d}u_z(x)}{\mathrm{d}x} . \tag{2.59}$$

The last relation equals $(\mathrm{d}s - \mathrm{d}x)$ in Eq. (2.52) and differentiation with respect to the x-coordinate leads directly to Eq. (2.56).

---

[12]Note that according to the assumptions of the Euler–Bernoulli beam the lengths 01 and $0'1'$ remain unchanged.

2.3.2 Constitutive Equation

The one-dimensional Hooke's law can be assumed in the case of the bending beam, since, according to the requirement, only normal stresses are regarded in this section:

$$\sigma_x = E\varepsilon_x \,. \tag{2.60}$$

where E is Young's modulus. Through the kinematics relation according to Eq. (2.56), the stress results as a function of the deflection to:

$$\sigma_x(x, z) = -Ez\frac{\mathrm{d}^2 u_z(x)}{\mathrm{d}x^2} \,. \tag{2.61}$$

The stress distribution shown in Fig. 2.19a generates the internal moment, which acts in this cross section. To calculate this internal moment, the stress is multiplied by a surface element, so that the resulting force is obtained. Multiplication with the corresponding lever arm then gives the internal moment. Since the stress is linearly distributed over the height, the evaluation is done for an infinitesimally small surface element:

$$\mathrm{d}M_y = (+z)(+\sigma_x)\mathrm{d}A = z\sigma_x\mathrm{d}A \,. \tag{2.62}$$

Therefore, the entire moment results via integration over the entire surface in:

$$M_y = \int_A z\sigma_x\mathrm{d}A \overset{(2.61)}{=} -\int_A zEz\frac{\mathrm{d}^2 u_z(x)}{\mathrm{d}x^2}\mathrm{d}A \,. \tag{2.63}$$

Assuming that the Young's modulus is constant, the internal moment around the y-axis results in:

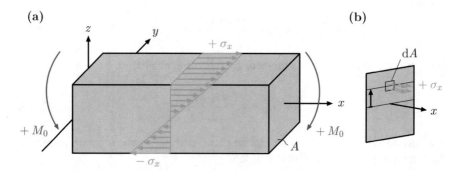

Fig. 2.19 **a** Schematic representation of the normal stress distribution $\sigma_x = \sigma_x(z)$ of a bending beam; **b** Definition and position of an infinitesimal surface element for the derivation of the resulting moment action due to the normal stress distribution

$$M_y = -E\frac{d^2 u_z}{dx^2} \underbrace{\int_A z^2 dA}_{I_y} = \frac{I_y \sigma_x}{z}. \tag{2.64}$$

The integral in Eq. (2.64) is the so-called axial second moment of area or axial surface moment of 2nd order in the SI unit m^4. This factor is only dependent on the geometry of the cross section and is also a measure of the stiffness of a plane cross section against bending. The values of the axial second moment of area for simple geometric cross sections are collected in Table 2.6.

Consequently the internal moment can also be expressed as

$$M_y = -EI_y \frac{d^2 u_z}{dx^2} \overset{(2.51)}{=} \frac{EI_y}{R} = EI_y \kappa. \tag{2.65}$$

Equation (2.65) describes the bending line $u_z(x)$ as a function of the bending moment and is therefore also referred to as the bending line-moment relation. The product EI_y in Eq. (2.65) is also called the bending stiffness. If the result from Eq. (2.65) is used in the relation for the bending stress according to Eq. (2.61), the distribution of stress over the cross section results in:

$$\sigma_x(x, z) = +\frac{M_y(x)}{I_y} z(x). \tag{2.66}$$

The plus sign in Eq. (2.66) causes that a positive bending moment (see Fig. 2.14) leads to a tensile stress in the upper beam half (meaning for $z > 0$).

In the case of plane bending with $M_y(x) \neq$ const., the bending line can be approximated in each case locally through a circle of curvature, see Fig. 2.20. Therefore, the result for *pure* bending according to Eq. (2.65) can be transferred to the case of plane bending as:

$$-EI_y \frac{d^2 u_z(x)}{dx^2} = M_y(x). \tag{2.67}$$

Let us note at the end of this section that Hooke's law in the form of Eq. (2.60) is not so easy to apply[13] in the case of beams since the stress and strain is linearly changing over the height of the cross section, see Eq. (2.66) and Fig. 2.19. Thus, it might be easier to apply a so-called stress resultant or generalized stress, i.e. a simplified representation of the normal stress state[14] based on the acting bending moment:

$$M_y(x) = \iint z\sigma_x(x, z)\, dA, \tag{2.68}$$

---

[13] However, this formulation works well in the case of rod elements since stress and strain are constant over the cross section, i.e. $\sigma_x = \sigma_x(x)$ and $\varepsilon_x = \varepsilon_x(x)$.

[14] A similar stress resultant can be stated for the shear stress based on the shear force: $Q_z(x) = \iint \tau_{xz}(x, z)\, dA$.

Table 2.6 Axial second moment of area around the y-axis

Cross section	I_y
	$\dfrac{\pi D^4}{64} = \dfrac{\pi R^4}{4}$
	$\dfrac{\pi b a^3}{4}$
	$\dfrac{a^4}{12}$
	$\dfrac{b h^3}{12}$
	$\dfrac{b h^3}{36}$
	$\dfrac{b h^3}{36}$

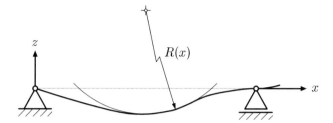

Fig. 2.20 Deformation of a beam in the *x*-*z* plane with $M_y(x) \neq$ const.

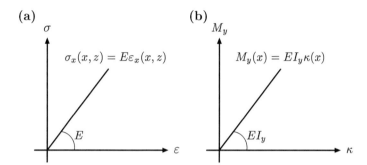

Fig. 2.21 Formulation of the constitutive law based on **a** stress and **b** stress resultant (bending in the *x*-*z* plane)

which was already introduced in Eq. (2.62). Using in addition the curvature[15] $\kappa = \kappa(x)$ (see Eq. (2.56)) instead of the strain $\varepsilon_x = \varepsilon_x(x, z)$, the constitutive equation can be easier expressed as shown in Fig. 2.21. The variables M_y and κ have both the advantage that they are constant for any location x of the beam.

2.3.3 Equilibrium

The equilibrium conditions are derived from an infinitesimal beam element of length dx, which is loaded by a constant distributed force $q_{z,0}$ as well as a constant distributed moment $m_{y,0}$, see Fig. 2.22. The internal reactions are drawn on both cut faces, i.e. at locations x and $x + dx$. One can see that a positive shear force is oriented in the positive *z*-direction at the right-hand face[16] and that a positive bending moment has

---

[15] The curvature is then called a generalized strain.

[16] A positive cut face is defined by the surface normal on the cut plane which has the same orientation as the positive *x*-axis. It should be regarded that the surface normal is always directed outward.

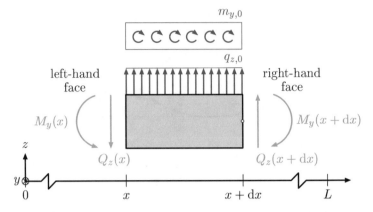

Fig. 2.22 Infinitesimal beam element in the x-z plane with internal reactions and constant distributed loads

the same rotational direction as the positive y-axis (right-hand grip rule[17]). The orientation of shear force and bending moment is reversed at the left-hand, i.e. negative, face in order to cancel in sum the effect of the internal reactions at both faces. This convention for the direction of the internal directions is maintained in the following. Furthermore, it can be derived from Fig. 2.22 that an upwards directed *external* force or alternatively a mathematically positive oriented *external* moment at the right-hand face leads to a positive internal shear force or alternatively a positive internal bending moment. In a corresponding way, it results that a downwards directed *external* force or alternatively a mathematically negative oriented *external* moment at the left-hand face leads to a positive internal shear force or alternatively a positive internal bending moment.

The equilibrium condition will be determined in the following for the vertical forces. Assuming that forces in the direction of the positive z-axis are considered positive, the following results:

$$- Q_z(x) + Q_z(x + \mathrm{d}x) + q_{z,0}\mathrm{d}x = 0 \,. \tag{2.69}$$

If the shear force on the right-hand face is expanded in a Taylor's series of first order, meaning

$$Q_z(x + \mathrm{d}x) \approx Q_z(x) + \frac{\mathrm{d}Q_z(x)}{\mathrm{d}x}\mathrm{d}x \,, \tag{2.70}$$

[17] If the axis is grasped with the right hand in a way so that the spread out thumb points in the direction of the positive axis, the bent fingers then show the direction of the positive rotational direction.

Equation (2.69) results in

$$- Q_z(x) + Q_z(x) + \frac{\mathrm{d}Q_z(x)}{\mathrm{d}x}\mathrm{d}x + q_{z,0}\mathrm{d}x = 0 \,, \qquad (2.71)$$

or alternatively after simplification finally to:

$$\frac{\mathrm{d}Q_z(x)}{\mathrm{d}x} = -q_{z,0} \,. \qquad (2.72)$$

For the special case that no distributed load is acting ($q_{z,0} = 0$), Eq. (2.72) simplifies to:

$$\frac{\mathrm{d}Q_z(x)}{\mathrm{d}x} = 0 \,. \qquad (2.73)$$

The equilibrium of moments around the reference point at $x + \mathrm{d}x$ gives:

$$M_y(x + \mathrm{d}x) - M_y(x) - Q_z(x)\mathrm{d}x + \frac{1}{2}q_{z,0}\mathrm{d}x^2 + m_{y,0}\mathrm{d}x = 0 \,. \qquad (2.74)$$

If the bending moment on the right-hand face is expanded into a Taylor's series of first order similar to Eq. (2.70) and consideration that the term $\frac{1}{2}q_{z,0}\mathrm{d}x^2$ as infinitesimal small size of higher order can be disregarded, finally the following results:

$$\frac{\mathrm{d}M_y(x)}{\mathrm{d}x} = Q_z(x) - m_{y,0} \,. \qquad (2.75)$$

The combination of Eqs. (2.72) and (2.75) leads to the relation between the bending moment and the distributed loads:

$$\frac{\mathrm{d}^2 M_y(x)}{\mathrm{d}x^2} = \frac{\mathrm{d}Q_z(x)}{\mathrm{d}x} - \frac{\mathrm{d}m_{y,0}}{\mathrm{d}x} = -q_{z,0} - \frac{\mathrm{d}m_{y,0}}{\mathrm{d}x} \,. \qquad (2.76)$$

Finally, the elementary basic equations for the bending of a beam in the *x*-*z* plane for arbitrary moment loading $M_y(x)$ are summarized in Table 2.7.

2.3.4 Differential Equation

Different formulations of the fourth-order differential equation are collected in Table 2.8 where different types of loadings, geometry and bedding are differentiated. The last case in Table 2.8 refers to the elastic foundation of a beam which is

Table 2.7 Elementary basic equations for the bending of a thin beam in the x-z plane. The differential equations are given under the assumption of constant bending stiffness EI_y

Name	Equation
Kinematics	$\varepsilon_x(x, z) = -z\dfrac{d^2u_z(x)}{dx^2}$
Equilibrium	$\dfrac{dQ_z(x)}{dx} = -q_z(x)\,;\ \dfrac{dM_y(x)}{dx} =$ $Q_z(x) - m_y(x)$
Constitutive equation	$\sigma_x(x, z) = E\varepsilon_x(x, z)$
Stress	$\sigma_x(x, z) = \dfrac{M_y(x)}{I_y}z(x)$
Diff'equation	$EI_y\dfrac{d^2u_z(x)}{dx^2} = -M_y(x)$
	$EI_y\dfrac{d^3u_z(x)}{dx^3} = -Q_z(x) + m_y(x)$
	$EI_y\dfrac{d^4u_z(x)}{dx^4} = q_z(x) + \dfrac{dm_y(x)}{dx}$

Table 2.8 Different formulations of the partial differential equation for a thin beam in the x-z plane (x-axis: right facing; z-axis: upward facing)

Configuration	Partial differential equation
E, I_y	$EI_y\dfrac{d^4u_z(x)}{dx^4} = 0$
$E(x), I_y(x)$	$\dfrac{d^2}{dx^2}\left(E(x)I_y(x)\dfrac{d^2u_z(x)}{dx^2}\right) = 0$
$q_z(x)$	$EI_y\dfrac{d^4u_z(x)}{dx^4} = q_z(x)$
$m_y(x)$	$EI_y\dfrac{d^4u_z(x)}{dx^4} = \dfrac{dm_y(x)}{dx}$
$k(x)$	$EI_y\dfrac{d^4u_z(x)}{dx^4} = -k(x)u_z(x)$

Table 2.9 Different formulations of the basic equations for a thin beam (bending in the x-z plane; x-axis along the principal beam axis). E: Young's modulus; I_y: second moment of area; q_z: length-specific distributed force; m_y: length-specific distributed moment; $\mathcal{L}_2 = \frac{d^2(...)}{dx^2}$: second-order derivative; $\mathcal{L}_1 = \frac{d(...)}{dx}$: first-order derivative

Specific formulation	General formulation [2]
Kinematics	
$\varepsilon_x(x,z) = -z\dfrac{d^2 u_z(x)}{dx^2}$	$\varepsilon_x(x,z) = -z\mathcal{L}_2\left(u_z(x)\right)$
$\kappa(x) = -\dfrac{d^2 u_z(x)}{dx^2}$	$\kappa(x) = -\mathcal{L}_2\left(u_z(x)\right)$
Constitution	
$\sigma_x(x,z) = E\varepsilon_x(x,z)$	$\sigma_x(x,z) = C\varepsilon_x(x,z)$
$M_y(x) = E I_y \kappa(x)$	$M_y(x) = D\kappa(x)$
Equilibrium	
$\dfrac{d^2 M_y(x)}{dx^2} + q_z(x) + \dfrac{dm_y(x)}{dx} = 0$	$\mathcal{L}_2^{\mathrm{T}}(M_y(x)) + q_z(x) + \mathcal{L}_1^{\mathrm{T}}(m_y(x)) = 0$
PDE	
$\dfrac{d^2}{dx^2}\left(E I_y \dfrac{d^2 u_z(x)}{dx^2}\right) - q_z(x) - \dfrac{dm_y(x)}{dx} = 0$	$\mathcal{L}_2^{\mathrm{T}}(D\mathcal{L}_2(u_z(x))) - q_z(x) - \mathcal{L}_1^{\mathrm{T}}(m_y(x)) = 0$

also know in the literature as Winkler foundation [15]. The elastic foundation or Winkler foundation modulus k has in the case of beams[18] the unit of force per unit area.

If we replace the common formulations of the second- and first-order derivatives, i.e. $\frac{d^2...}{dx^2}$ and $\frac{d...}{dx}$, by formal operator symbols, i.e. $\mathcal{L}_2(...)$ and $\mathcal{L}_1(...)$, the basic equations can be stated in a more formal way as given in Table 2.9. Such a general formulation is suitable for the derivation of the principal finite element equation based on the weighted residual method [10, 11].

Under the assumption of constant material ($E = $ const.) and geometric ($I_y = $ const.) properties, the differential equation in Table 2.9 can be integrated four times for constant distributed loads ($q_z(x) = q_0 = $ const., $m_y(x) = 0$) to obtain the general analytical solution of the problem:

$$u_z(x) = \frac{1}{E I_y}\left(\frac{q_0 x^4}{24} + \frac{c_1 x^3}{6} + \frac{c_2 x^2}{2} + c_3 x + c_4\right), \qquad (2.77)$$

where the four constants of integration c_i $(i = 1, \ldots, 4)$ must be determined based on the boundary conditions. The following equations for the shear force $Q_z(x)$, the bending moment $M_y(x)$, and the rotation $\varphi_y(x)$ were obtained based on one-, two-

---

[18] In the general case, the unit of the elastic foundation modulus is force per unit area per unit length, i.e. $\frac{N}{m^2}/m = \frac{N}{m^3}$.

and three-times integration and might be useful to determine some of the constants of integration:

$$Q_z(x) = -q_0 x - c_1, \tag{2.78}$$

$$M_y(x) = -\frac{q_0 x^2}{2} - c_1 x - c_2, \tag{2.79}$$

$$\varphi_y(x) = -\frac{du_z(x)}{dx} = -\frac{1}{EI_y}\left(\frac{q_0 x^3}{6} + \frac{c_1 x^2}{2} + c_2 x + c_3\right). \tag{2.80}$$

2.4 Comparison of Both Planes

The comparison of the elementary basic equations for bending in the x-y and x-y plane are shown in Table 2.10. All the basic equations have the same structure but the different definition of a positive rotation yields to some different signs.

Table 2.10 Elementary basic equations for the bending of a thin beam in the x-y and x-y plane. The differential equations are given under the assumption of constant bending stiffness EI

Equation	x-y plane	x-z plane
Kinematics	$\varepsilon_x(x, y) = -y\dfrac{d^2 u_y(x)}{dx^2}$	$\varepsilon_x(x, z) = -z\dfrac{d^2 u_z(x)}{dx^2}$
Equilibrium	$\dfrac{dQ_y(x)}{dx} = -q_y(x)$	$\dfrac{dQ_z(x)}{dx} = -q_z(x)$
	$\dfrac{dM_z(x)}{dx} = -Q_y(x) - m_z(x)$	$\dfrac{dM_y(x)}{dx} = Q_z(x) - m_y(x)$
Constitution	$\sigma_x(x, y) = E\varepsilon_x(x, y)$	$\sigma_x(x, z) = E\varepsilon_x(x, z)$
	$M_z(x) = EI_z\kappa(x)$	$M_y(x) = EI_y\kappa(x)$
Stress	$\sigma_x(x, y) = -\dfrac{M_z(x)}{I_z}y(x)$	$\sigma_x(x, z) = \dfrac{M_y(x)}{I_y}z(x)$
Diff'equation	$EI_z\dfrac{d^2 u_y(x)}{dx^2} = M_z(x)$	$EI_y\dfrac{d^2 u_z(x)}{dx^2} = -M_y(x)$
	$EI_z\dfrac{d^3 u_y(x)}{dx^3} = -Q_y(x) - m_z(x)$	$EI_y\dfrac{d^3 u_z(x)}{dx^3} = -Q_z(x) + m_y(x)$
	$EI_z\dfrac{d^4 u_y(x)}{dx^4} = q_y(x) - \dfrac{dm_z(x)}{dx}$	$EI_y\dfrac{d^4 u_z(x)}{dx^4} = q_z(x) + \dfrac{dm_y(x)}{dx}$

2.5   Unsymmetrical Bending in Both Planes

Depending on the cross section and the position of the moment vector in regards to the principal coordinate axes, i.e., a coordinate system where the mixed second moment of area is zero, symmetrical or unsymmetrical bending can occur, see Fig. 2.23.

Let us assume in the following that the origin of a Cartesian y-z coordinate system is located in the centroid of an arbitrary cross-section beam, see Fig. 2.24a.

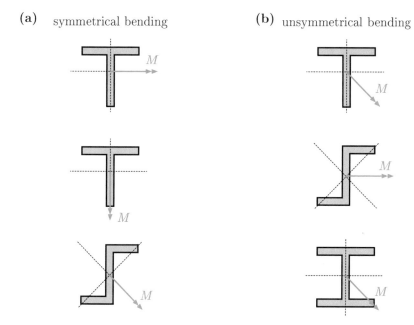

Fig. 2.23   a Symmetrical bending and **b** unsymmetrical bending (the principal axis system is represented by dashed lines)

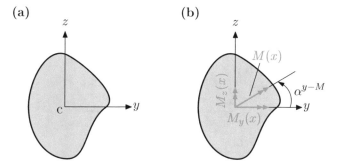

Fig. 2.24   Arbitrary cross-section beam: **a** Origin of the y-z coordinate system coincides with the centroid ('c') of the beam. **b** Decomposition of the internal bending moment in its components

The acting internal bending moment $M(x)$ can be split into its components acting around the y- and z-axis based on the following transformations, see Fig. 2.24b:

$$M_y(x) = |M(x)| \cos(\alpha^{y-M}(x)),$$ (2.81)
$$M_z(x) = |M(x)| \sin(\alpha^{y-M}(x)).$$ (2.82)

The total deflection of the beam results from the components based on vector summation as:

$$M(x) = \sqrt{M_y(x)^2 + M_z(x)^2}.$$ (2.83)

In addition, the angle of rotation between the y-axis and the bending moment $M(x)$ can be expressed as:

$$\tan(\alpha^{y-M}(x)) = \frac{M_z(x)}{M_y(x)}.$$ (2.84)

The total normal strain in the x-direction is obtained by simple superposition of its components from the bending contributions in the x-y and x-z plane (see Eqs. (2.16) and (2.56) for details), i.e.

$$\varepsilon_x(x, y, z) = \varepsilon_x^{x-y}(x, y) + \varepsilon_x^{x-z}(x, z).$$ (2.85)

Thus, Hooke's law according to Eqs. (2.20) or (2.60) can be generalized to the following formulation:

$$\sigma_x(x, y, z) = E\varepsilon_x(x, y, z) = -E\left(y\frac{d^2 u_y(x)}{dx^2} + z\frac{d^2 u_z(x)}{dx^2}\right).$$ (2.86)

Based on this stress distribution, the evaluation of the internal bending moments $M_z(x)$ and $M_y(x)$ (see Eqs. (2.23) and (2.63) for details) can be performed to obtain the following relations:

$$M_z(x) = -\int_A y\sigma_x dA = +E\int_A\left(y^2\frac{d^2 u_y(x)}{dx^2} + yz\frac{d^2 u_z(x)}{dx^2}\right)dA$$ (2.87)

$$= +E\frac{d^2 u_y(x)}{dx^2}\underbrace{\int_A y^2 dA}_{I_z} + E\frac{d^2 u_z(x)}{dx^2}\underbrace{\int_A yz dA}_{-I_{yz}}$$ (2.88)

$$= +EI_z\frac{d^2 u_y(x)}{dx^2} - EI_{yz}\frac{d^2 u_z(x)}{dx^2},$$ (2.89)

or for the internal bending moment around the y-axis:

$$M_y(x) = + \int_A z\sigma_x dA = -E \int_A \left(yz\frac{d^2u_y(x)}{dx^2} + z^2\frac{d^2u_z(x)}{dx^2} \right) dA \qquad (2.90)$$

$$= -E\frac{d^2u_y(x)}{dx^2}\underbrace{\int_A yz dA}_{-I_{yz}} - E\frac{d^2u_z(x)}{dx^2}\underbrace{\int_A z^2 dA}_{I_y} \qquad (2.91)$$

$$= +EI_{yz}\frac{d^2u_y(x)}{dx^2} - EI_y\frac{d^2u_z(x)}{dx^2}. \qquad (2.92)$$

The last two relations for the internal bending moment can be rearranged for the derivative of the displacement field. From Eq. (2.89), one can obtain

$$\frac{d^2u_z(x)}{dx^2} = -\frac{M_z(x) - EI_z\frac{d^2u_y(x)}{dx^2}}{EI_{yz}}, \qquad (2.93)$$

which can be introduced into Eq. (2.92) to get:

$$\frac{d^2u_y(x)}{dx^2} = \frac{I_y M_z(x) - I_{yz} M_y(x)}{E\left(I_y I_z - I_{yz}^2\right)}. \qquad (2.94)$$

The last formulation can be introduced into Eq. (2.89) to finally give:

$$\frac{d^2u_z(x)}{dx^2} = -\frac{I_z M_y(x) - I_{yz} M_z(x)}{E\left(I_y I_z - I_{yz}^2\right)}. \qquad (2.95)$$

Thus, the deformations of an unsymmetrical bending problem can be obtained based on the decoupled second-order differential equations (2.94) and (2.95). Once the deformation components $u_y(x)$ and $u_z(x)$ are know, a vector addition allows to calculate the total deformation as follows:

$$u(x) = \sqrt{(u_y(x))^2 + (u_y(x))^2}. \qquad (2.96)$$

It should be noted here that the differential equations (2.94) and (2.95) reduce for a principal axis system, i.e., $I_{yz} = 0$, to the formulations provided in Tables 2.3 and 2.7.

   An alternative formulations of the partial differential equations can be obtained by considering the equilibrium equations (2.36) and (2.76). Combining these equilibrium equations with the moment relations of Eqs. (2.89) and (2.92), gives finally the following set of coupled fourth-order differential equations based on the distributed load components:

$$\frac{d^2}{dx^2}\left(+EI_z\frac{d^2u_y(x)}{dx^2} - EI_{yz}\frac{d^2u_z(x)}{dx^2}\right) = q_y(x) - \frac{dm_z(x)}{dx}, \qquad (2.97)$$

$$\frac{d^2}{dx^2}\left(+EI_{yz}\frac{d^2u_y(x)}{dx^2} - EI_y\frac{d^2u_z(x)}{dx^2}\right) = -q_z(x) - \frac{dm_y(x)}{dx}, \qquad (2.98)$$

or in matrix notation:

$$\frac{d^2}{dx^2}\left(\begin{bmatrix} I_z & -I_{yz} \\ I_{yz} & -I_y \end{bmatrix}\begin{bmatrix} \dfrac{d^2u_y(x)}{dx^2} \\ \dfrac{d^2u_z(x)}{dx^2} \end{bmatrix}\right) = \begin{bmatrix} q_y(x) - \dfrac{dm_z(x)}{dx} \\ -q_z(x) - \dfrac{dm_y(x)}{dx} \end{bmatrix}. \qquad (2.99)$$

Let us focus in the following on the stress distribution, which was originally given in Eq. (2.86). Introducing the relations for the second-order derivatives of the displacement fields as given in Eqs. (2.94) and (2.95) gives (see Fig. 2.25 for a graphical illustration):

$$\sigma_x(x, y, z) = \frac{\left(I_z M_y(x) - I_{yz} M_z(x)\right)z - \left(I_y M_z(x) - I_{yz} M_y(x)\right)y}{I_y I_z - I_{yz}^2}. \qquad (2.100)$$

The functional equation of the neutral fiber (index 'n') results from Eq. (2.100) based on the condition $\sigma_x = 0$ as:

$$z_n(y_n) = \frac{I_y M_z(x) - I_{yz} M_y(x)}{I_z M_y(x) - I_{yz} M_z(x)} \times y_n. \qquad (2.101)$$

The corresponding slope allows to calculate the rotational angle β^{y-n} between the y-axis and the neutral fiber (counterclockwise positive):

$$\frac{dz_n(y_n)}{dy_n} = \tan \beta^{y-n} = \frac{I_y M_z(x) - I_{yz} M_y(x)}{I_z M_y(x) - I_{yz} M_z(x)}. \qquad (2.102)$$

Let us consider in the following a principal coordinate system (η, ζ) for which the mixed second moment of area is zero ($I_{\eta\zeta} = 0$), see Fig. 2.26. This principal coordinate system is obtained from the original y-z coordinate system by a pure rotation of angle $\alpha^{y-\eta}$.

Fig. 2.25 Stress distribution (special case of doubly symmetric cross section): **a** only $+M_y$ acting, **b** only $+M_z$ acting and **c** $+M_y$ and $+M_z$ acting

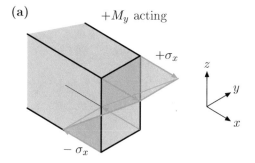

(a) $+M_y$ acting

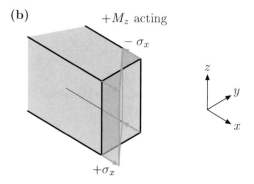

(b) $+M_z$ acting

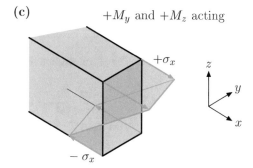

(c) $+M_y$ and $+M_z$ acting

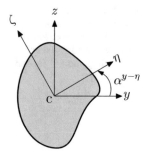

Fig. 2.26 Arbitrary cross-section beam. y-z: arbitrary Cartesian coordinate system ($I_{xy} \neq 0$). η-ζ: principal coordinate system ($I_{\eta\zeta} = 0$)

$$I_\eta = \frac{I_y + I_z}{2} + \sqrt{\left(\frac{I_y - I_z}{2}\right)^2 + I_{yz}^2}\,, \qquad (2.103)$$

$$I_\zeta = \frac{I_y + I_z}{2} - \sqrt{\left(\frac{I_y - I_z}{2}\right)^2 + I_{yz}^2}\,. \qquad (2.104)$$

The angle of rotation can be calculated based on the second moments of area in the arbitrary x-y system as:

$$\alpha^{y-\eta,\zeta} = \frac{1}{2} \times \arctan\left(\frac{2I_{yz}}{I_y - I_z}\right)\,, \qquad (2.105)$$

whereas the second-order derivative of Eq. (B.8), i.e.,

$$\frac{\mathrm{d}^2 I_\eta(\alpha^{y-\eta,\zeta})}{\mathrm{d}(\alpha^{y-\eta,\zeta})^2} = -2\left(I_y - I_z\right)\cos(2\alpha^{y-\eta,\zeta}) - 4I_{yz}\sin(2\alpha^{y-\eta,\zeta})\,, \qquad (2.106)$$

allows to decide if $\alpha^{y-\eta,\zeta}$ is the angle to the η- or to the ζ-axis:

$$\begin{aligned} \frac{\mathrm{d}^2 I_\eta(\alpha^{y-\eta,\zeta})}{\mathrm{d}(\alpha^{y-\eta,\zeta})^2} &< 0 \;\rightarrow\; \alpha = \alpha^{y-\eta}\,, \\ \frac{\mathrm{d}^2 I_\eta(\alpha^{y-\eta,\zeta})}{\mathrm{d}(\alpha^{y-\eta,\zeta})^2} &> 0 \;\rightarrow\; \alpha = \alpha^{y-\zeta}\,. \end{aligned} \qquad (2.107)$$

Considering the condition $I_{\eta\zeta} = 0$, the fourth-order differential equations (2.97) and (2.98) can be simplified for the η-ζ system to the following formulation:

$$+EI_\zeta\frac{\mathrm{d}^4 u_\eta(x)}{\mathrm{d}x^4} = q_\eta(x) - \frac{\mathrm{d}m_\zeta(x)}{\mathrm{d}x}\,, \qquad (2.108)$$

$$-EI_\eta\frac{\mathrm{d}^4 u_\zeta(x)}{\mathrm{d}x^4} = -q_\zeta(x) - \frac{\mathrm{d}m_\eta(x)}{\mathrm{d}x}\,, \qquad (2.109)$$

Fig. 2.27 Position of the bending moment and its components in regards to the principal coordinate axis ($I_{\eta\zeta} = 0$)

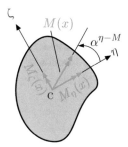

or in matrix notation:

$$\frac{d^2}{dx^2}\left(\begin{bmatrix} I_\zeta & 0 \\ 0 & -I_\eta \end{bmatrix} \begin{bmatrix} \dfrac{d^2 u_\eta(x)}{dx^2} \\ \dfrac{d^2 u_\zeta(x)}{dx^2} \end{bmatrix} \right) = \begin{bmatrix} q_\eta(x) - \dfrac{dm_\zeta(x)}{dx} \\ -q_\zeta(x) - \dfrac{dm_\eta(x)}{dx} \end{bmatrix}. \tag{2.110}$$

Correspondingly, the stress distribution (2.100) can be simplified to

$$\sigma_x(x, \eta, \zeta) = \frac{M_\eta(x)}{I_\eta}\zeta - \frac{M_\zeta(x)}{I_\zeta}\eta, \tag{2.111}$$

whereas the neutral fiber (2.101) is now obtained as:

$$\zeta_n(\eta_n) = \frac{I_\eta M_\zeta(x)}{I_\zeta M_\eta(x)} \times \eta_n. \tag{2.112}$$

The corresponding slope allows to calculate the rotational angle $\beta^{\eta-n}$ between the η-axis and the neutral fiber (counterclockwise positive):

$$\frac{d\zeta_n(\eta_n)}{d\eta_n} = \tan(\beta^{\eta-n}) = \frac{I_\eta M_\zeta(x)}{I_\zeta M_\eta(x)}. \tag{2.113}$$

The components of the bending moment in Eq. (2.111), i.e., $M_\eta(x)$ and $M_\zeta(x)$ (see Fig. 2.27), can be obtained from the total moment $M(x)$ (see also Eq. (2.83)):

$$M_\eta(x) = |M(x)|\cos(\alpha^{\eta-M}(x)), \tag{2.114}$$

$$M_\zeta(x) = |M(x)|\sin(\alpha^{\eta-M}(x)). \tag{2.115}$$

Thus, Eq. (2.111) can be alternatively written as follows:

$$\sigma_x(x, \eta, \zeta) = |M(x)|\left(\frac{\cos(\alpha^{\eta-M}(x))}{I_\eta}\zeta - \frac{\sin(\alpha^{\eta-M}(x))}{I_\zeta}\eta \right). \tag{2.116}$$

The η-ζ coordinates in Eq. (2.116) are sometimes easier expressed in a y-z system (e.g., in the case that a coordinate axis is aligned to an edge of the cross-sectional profile). The transformation of a point A given in the y-z coordinate system (y_A, z_A) into the η-ζ coordinate system can be based on the following rotational transformations:

$$\eta_A = y_A \cos(-\alpha^{y-\eta}) - z_A \sin(-\alpha^{y-\eta}), \qquad (2.117)$$

$$\zeta_A = y_A \sin(-\alpha^{y-\eta}) + z_A \cos(-\alpha^{y-\eta}). \qquad (2.118)$$

Let us summarize at the end of this section the recommended steps for a stress analysis in the case of unsymmetrical bending problems ('hand calculation'):

① Calculate the center of area ('centroid') of the cross section (see 'c' in Fig. 2.24a). If a cross section is composed of basic surfaces whose centroidal locations are known, Eqs. (B.3) and (B.4) can be applied.

② Introduce the Cartesian y-x coordinate system. Locate the origin in the centroid (see Fig. 2.24a) and align axes, if possible, along straight edges of the cross section. The x-axis is pointing out of your paper/screen, i.e., a positive rotation in the y-x plane is counterclockwise.

③ Calculate the second moments of area in the y-x coordinate system, i.e., I_y, I_z, I_{yz}.

④ Calculate the *principal* second moments of area I_η, I_ζ (see Eqs. (2.103) and (2.104)) and the angle of rotation $\alpha^{y-\eta}$ between the y- and the η-axis (see Eq. (2.105)).

⑤ Calculate the equation of the neutral fiber $\zeta_n = \zeta_n(\eta_n)$ in the η-ζ coordinate system (see Eq. (2.112)). Identify on both sides of the neutral axis the point on the surface with the maximum distance to the neutral fiber (critical stress points).

⑥ Calculate the normal stresses in the critical stress points based on Eq. (2.116). The coordinates of the critical stress points must be expressed in the η-ζ coordinate system. Equations (2.117) and (2.118) allow to derive these coordinates based on the coordinates in the y-z system.

2.6 Generalized Beam Formulation: Consideration of Tension and Compression

The simple superposition of a pure bending beam with a tensile rod is considered, see Fig. 2.28. Let us assume in the following that the origin of the y-z coordinate system is located in the centroid of the surface.

The elementary basic equations for the simple superposition of a bending beam and a tensile bar in the x-y plane are summarized in Table 2.11. It is assumed for simplicity that the material (E) and geometrical properties (I_z, A) are constant.

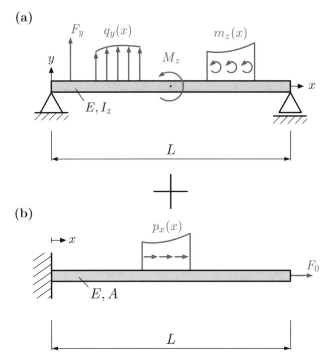

Fig. 2.28 Simple superposition of a bending beam and a tensile bar in the x-y plane: **a** pure bending deformation; **b** pure elongation

The total displacement field is obtained by means of vector addition of the single components originating from the bending and from the tensile mode of deformation:

$$u(x) = \sqrt{(u_x(x))^2 + (u_y(x))^2}\,. \tag{2.119}$$

The different stress distributions of the normal stress σ_x, i.e., linear for the beam and constant for the tensile bar, and the corresponding superposition are illustrated in Fig. 2.29 for the x-y plane. It can be seen that an unsymmetrical linear distribution is obtained whose maximum is located at the lower side of the beam.

The corresponding simple superposition of a pure bending beam in the x-z plane with a tensile rod is shown in Fig. 2.30.

Similar to the summary previous summary, Table 2.12 collects the elementary basic equations for the simple superposition of a bending beam and a tensile bar in the x-z plane. Obviously that the tensile part remains the same as shown in Table 2.11.

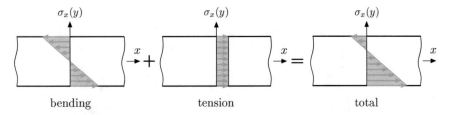

Fig. 2.29 Stress distributions for the simple superposition of a bending beam (symmetrical cross section assumed) and a tensile bar in the x-y plane **a** pure bending; **b** pure tension; **c** superposition of both cases

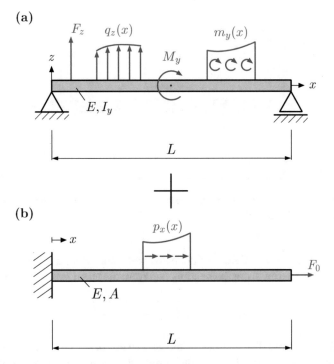

Fig. 2.30 Simple superposition of a bending beam and a tensile bar in the x-z plane: **a** pure bending deformation; **b** pure elongation

The total displacement field for bending in the x-z plane together with a tensile deformation is again obtained by means of vector addition of the single components originating from the bending and from the tensile mode of deformation:

$$u(x) = \sqrt{(u_x(x))^2 + (u_z(x))^2}\,. \tag{2.120}$$

Table 2.11 Elementary basic equations for the simple superposition of a bending beam and a tensile bar in the x-y plane. The differential equations are given under the assumption of constant material and geometrical properties

Equation	Bending	Tension
Kinematics	$\varepsilon_x(x, y) = -y\dfrac{d^2 u_y(x)}{dx^2}$	$\varepsilon_x(x) = \dfrac{du_x(x)}{dx}$
Equilibrium	$\dfrac{dQ_y(x)}{dx} = -q_y(x)$ $\dfrac{dM_z(x)}{dx} = -Q_y(x) - m_z(x)$	$\dfrac{dN_x(x)}{dx} = -p_x(x)$
Constitution	$\sigma_x(x, y) = E\varepsilon_x(x, y)$	$\sigma_x(x) = E\varepsilon_x(x)$
Stress	$\sigma_x(x, y) = -\dfrac{M_z(x)}{I_z}y(x)$	$\sigma_x(x) = \dfrac{N_x(x)}{A}$
Diff'equation	$EI_z\dfrac{d^2 u_y(x)}{dx^2} = M_z(x)$ $EI_z\dfrac{d^3 u_y(x)}{dx^3} = -Q_y(x) - m_z(x)$ $EI_z\dfrac{d^4 u_y(x)}{dx^4} = q_y(x) - \dfrac{dm_z(x)}{dx}$	$EA\dfrac{du_x(x)}{dx} = N_x(x)$ $EA\dfrac{d^2 u_x(x)}{dx^2} = -p_x(x)$

Table 2.12 Elementary basic equations for the simple superposition of a bending beam and a tensile bar in the x-z plane. The differential equations are given under the assumption of constant material and geometrical properties

Equation	Bending	Tension
Kinematics	$\varepsilon_x(x, z) = -z\dfrac{d^2 u_z(x)}{dx^2}$	$\varepsilon_x(x) = \dfrac{du_x(x)}{dx}$
Equilibrium	$\dfrac{dQ_z(x)}{dx} = -q_z(x)$ $\dfrac{dM_y(x)}{dx} = Q_z(x) - m_y(x)$	$\dfrac{dN_x(x)}{dx} = -p_x(x)$
Constitution	$\sigma_x(x, z) = E\varepsilon_x(x, z)$	$\sigma_x(x) = E\varepsilon_x(x)$
Stress	$\sigma_x(x, z) = \dfrac{M_y(x)}{I_y}z(x)$	$\sigma_x(x) = \dfrac{N_x(x)}{A(x)}$
Diff'equation	$EI_y\dfrac{d^2 u_z(x)}{dx^2} = -M_y(x)$ $EI_y\dfrac{d^3 u_z(x)}{dx^3} = -Q_z(x) + m_y(x)$ $EI_y\dfrac{d^4 u_z(x)}{dx^4} = q_z(x) + \dfrac{dm_y(x)}{dx}$	$EA\dfrac{du_x(x)}{dx} = N_x(x)$ $EA\dfrac{d^2 u_x(x)}{dx^2} = -p_x(x)$

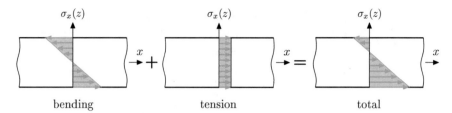

Fig. 2.31 Stress distributions for the simple superposition of a bending beam (symmetrical cross section assumed) and a tensile bar in the x-z plane: **a** pure bending; **b** pure tension; **c** superposition of both cases

The different stress distributions of the normal stress σ_x, i.e., linear for the beam and constant for the tensile bar, and the corresponding superposition are illustrated in Fig. 2.31 for the x-z plane. It can be seen again that an unsymmetrical linear distribution is obtained whose maximum is located at the lower side of the beam.

As introduced in Sect. 2.5, one can combine the unsymmetrical bending in two planes with the tensile mode to obtain the following system of differential equations:

$$\frac{d^2}{dx^2}\left(\begin{bmatrix} EA & 0 & 0 \\ 0 & I_z & -I_{yz} \\ 0 & I_{yz} & -I_y \end{bmatrix}\begin{bmatrix} \dfrac{du_x(x)}{dx} \\ \dfrac{d^2u_y(x)}{dx^2} \\ \dfrac{d^2u_z(x)}{dx^2} \end{bmatrix}\right) = \begin{bmatrix} -\dfrac{dp_x(x)}{dx} \\ q_y(x) - \dfrac{dm_z(x)}{dx} \\ -q_z(x) - \dfrac{dm_y(x)}{dx} \end{bmatrix}. \qquad (2.121)$$

2.7 Shear Stress Distribution Due to Transverse Loads

2.7.1 Solid Sections Beams

Starting point for the derivation of the shear stress distribution is, for example, the cantilever beam with a transverse force as shown in Fig. 2.32. Furthermore, only rectangular cross sections of dimension $b \times h$ will be considered in the following.

An infinitesimal beam element (in horizontal direction) of this configuration is shown in Fig. 2.33. The internal reactions are drawn according the sign convention in Fig. 2.22. Under the assumption that no distributed loads are acting (i.e., $q_z = 0$ and $m_y = 0$), it follows from the vertical force equilibrium that $Q_z(x) \approx Q_z(x + dx)$.

The next step is to replace the internal reactions, i.e. the bending moment and the shear force, by the corresponding normal and shear stresses. To this end, a part of the infinitesimal (in horizontal direction) beam element is cut off at $z = z'$, see Fig. 2.34. The vertical dimension of this element is now $h/2 - z'$.

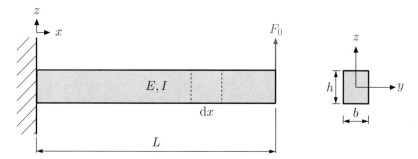

Fig. 2.32 General configuration of a cantilever beam with transverse force

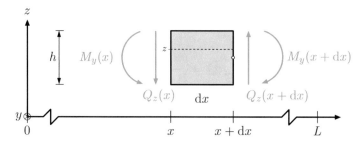

Fig. 2.33 Infinitesimal beam element $\mathrm{d}x$ in the x-z plane with internal reactions

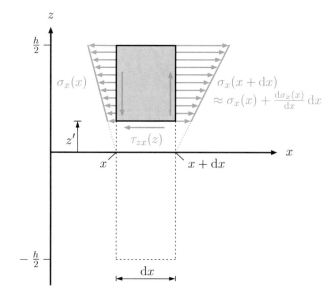

Fig. 2.34 Infinitesimal beam element of size $\mathrm{d}x \times (h/2 - z')$. The general configuration is shown in Fig. 2.32

The force equilibrium for this infinitesimal element in the x-direction gives:

$$- \sigma_x(x) \underbrace{b(z)\mathrm{d}z}_{\mathrm{d}A} + \sigma_x(x + \mathrm{d}x) \underbrace{b(z)\mathrm{d}z}_{\mathrm{d}A} - \tau_{zx}(z)b(z)\mathrm{d}x = 0 \,, \qquad (2.122)$$

or simplified after a Taylor's series expansion of the normal stress at $(x + \mathrm{d}x)$:

$$\frac{\mathrm{d}\sigma_x(x)}{\mathrm{d}x} \mathrm{d}x\mathrm{d}A - \tau_{zx}(z)b(x)\mathrm{d}x = 0 \,, \qquad (2.123)$$

or

$$\frac{\mathrm{d}\sigma_x(x)}{\mathrm{d}x} \mathrm{d}A - \tau_{zx}(z)b(x) = 0 \,. \qquad (2.124)$$

Rearranged for

$$\tau_{zx}(z) = \frac{\mathrm{d}\sigma_x(x)}{b(z)\mathrm{d}x} \mathrm{d}A \,, \qquad (2.125)$$

it follows with

$$\frac{\mathrm{d}\sigma_x(x)}{\mathrm{d}x} = \frac{\mathrm{d}}{\mathrm{d}x}\left(\frac{M_y(x)}{I_y} \times z\right) = \frac{z}{I_y} \times \frac{\mathrm{d}M_y(x)}{\mathrm{d}x} = \frac{Q_z(x)}{I_y} \times z \qquad (2.126)$$

for the shear stress distribution:

$$\tau_{zx}(z) = \frac{1}{b(z)} \int \frac{Q_z(x)}{I_y} \times z'\mathrm{d}A = \frac{Q_z(x)}{I_y b(z)} \int z'\mathrm{d}A = \frac{Q_z(x)\mathcal{H}_y(z)}{I_y b(z)} \,, \qquad (2.127)$$

where $\mathcal{H}_y(z)$ is the first moment of area for the part of the cross section shown in Fig. 2.34. Under the assumption of a rectangular cross section of dimension $b \times h$, this moment reads

$$\mathcal{H}_y(z) = \int z'\mathrm{d}A = b \int_z^{h/2} z'\mathrm{d}z' = \frac{b}{2}\left(\frac{h^2}{4} - z^2\right) \qquad (2.128)$$

$$= \frac{bh^2}{8}\left[1 - \left(\frac{z}{h/2}\right)^2\right] \,. \qquad (2.129)$$

Thus, the shear stress distribution in a beam with a rectangular cross section ($b \times h$) is finally obtained under consideration of $\tau_{zx} = \tau_{xz}$ as follows:

$$\tau_{xz}(z) = \frac{Q_z(x)h^2}{8I_y}\left[1 - \left(\frac{z}{h/2}\right)^2\right] \tag{2.130}$$

$$= \frac{3Q_z(x)}{2A}\left[1 - \left(\frac{z}{h/2}\right)^2\right]. \tag{2.131}$$

The maximum shear stress is obtained for $z = 0$:

$$\tau_{xz,\mathrm{max}} = \frac{Q_z(x)h^2}{8I_y} = \frac{3Q_z(x)}{2bh} = \frac{3Q_z(x)}{2A}. \tag{2.132}$$

2.7.2   Thin-Walled Beams

Let us focus now on thin-walled beams such as the I-profile shown in Fig. 2.35. Such a cross section is characterized by the fact that the thickness of the flange (t_f) and of the web (t_w) are small compared to the outer width (b) and height ($2c$). These geometrical constraints result in the following simplifications:

- The shear stress is parallel to the outer boundary of the profile.
- The shear stress is constant over the thickness.
- The vertical shear stress in the flanges and the horizontal shear stress in the web are neglected.

The first step is the calculation of the shear stress distribution in the flange (index 'f'). To do so, let us consider an infinitesimal flange element as shown in Fig. 2.36. The force equilibrium for this infinitesimal element in the x-direction gives:

$$\sigma_x(x + \mathrm{d}x, z)\mathrm{d}A - \sigma_x(x, z)\mathrm{d}A - \tau_{lf}(\eta)t_f\mathrm{d}x = 0, \tag{2.133}$$

Fig. 2.35 I-profile for the calculation of the shear stress distribution

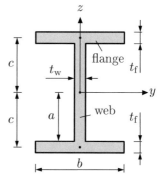

(a)

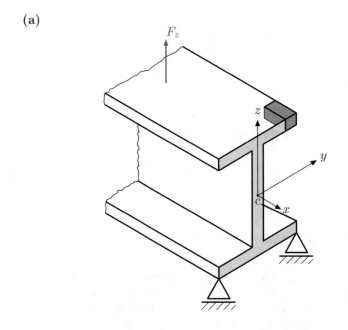

(b)

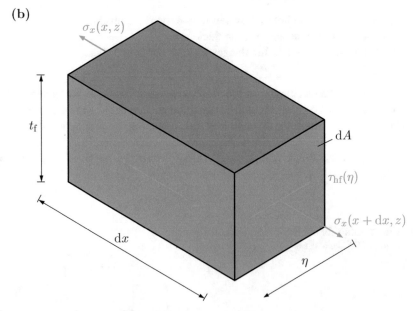

Fig. 2.36 Calculation of the shear stress in the flange: **a** general configuration and **b** infinitesimal flange element

or simplified after a Taylor's series expansion of the normal stress at $(x + dx)$:

$$\tau_{\mathrm{lf}}(\eta) = \frac{1}{t_{\mathrm{f}}} \times \frac{d\sigma_x}{dx} dA . \tag{2.134}$$

It follows with (see dimension c in Fig. 2.35)

$$\frac{d\sigma_x(z = c)}{dx} = \frac{d}{dx} \left(\frac{M_y(x)}{I_y} \times c \right) = \frac{c}{I_y} \times \frac{dM_y(x)}{dx} = \frac{Q_z(x)c}{I_y} \tag{2.135}$$

and the identity $\tau_{\mathrm{lf}}(\eta) = \tau_{\mathrm{hf}}(\eta)$ (index 'l': longitudinal; index 'h': horizontal) the following expression for the horizontal shear stress in the flange of a thin-walled I-profile:

$$\tau_{\mathrm{hf}}(\eta) = \frac{Q_z(x)c}{I_y t_{\mathrm{f}}} \int dA = \frac{Q_z(x)c}{I_y t_{\mathrm{f}}} \times t_{\mathrm{f}}\eta = \frac{Q_z(x)c}{I_y} \times \eta . \tag{2.136}$$

The next step is the calculation of the shear stress distribution in the web (index 'w'). To do so, let us consider the part of the web as shown in Fig. 2.37.

The force equilibrium for this part in the x-direction gives $(0 \le z \le a)$:

$$\sigma_x(x + dx, z)dA - \sigma_x(x, z)dA - \tau_{\mathrm{lw}}(\eta)t_{\mathrm{w}}dx = 0 , \tag{2.137}$$

or simplified after a Taylor's series expansion of the normal stress at $(x + dx)$:

$$\tau_{\mathrm{lw}}(z) = \frac{1}{t_{\mathrm{w}}} \times \frac{d\sigma_x}{dx} dA . \tag{2.138}$$

It follows with

$$\frac{d\sigma_x(z = \zeta)}{dx} = \frac{d}{dx} \left(\frac{M_y(x)}{I_y} \times \zeta \right) = \frac{\zeta}{I_y} \times \frac{dM_y(x)}{dx} = \frac{Q_z(x)\zeta}{I_y} \tag{2.139}$$

and the identity $\tau_{\mathrm{lw}}(z) = \tau_{\mathrm{vf}}(z)$ (index 'l': longitudinal; index 'v': vertical) the following expression for the vertical shear stress in the web of a thin-walled I-profile:

$$\tau_{\mathrm{vw}}(z) = \frac{Q_y(x)}{I_y t_{\mathrm{w}}} \underbrace{\int \zeta dA}_{\mathcal{H}_y(z)} . \tag{2.140}$$

The first moment of area $\mathcal{H}_y(z)$ for the part of the cross section shown in Fig. 2.37b reads in this case:

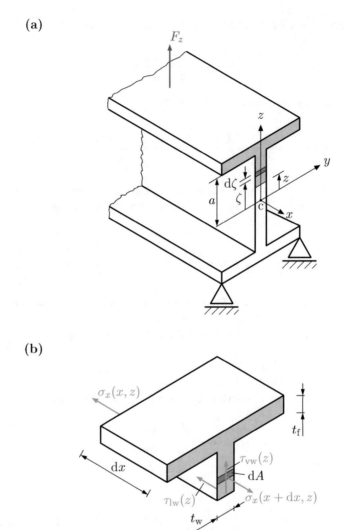

Fig. 2.37 Calculation of the shear stress in the web: **a** general configuration and **b** part of the web

$$\mathcal{H}_y(z) = \int \zeta dA = \int\limits_{\zeta=a}^{\zeta=a+t_f} \int\limits_{y=-\frac{b}{2}}^{y=+\frac{b}{2}} \zeta dy d\zeta + \int\limits_{\zeta=z}^{\zeta=a} \int\limits_{y=-\frac{t_w}{2}}^{y=-\frac{t_w}{2}} \zeta dy d\zeta \qquad (2.141)$$

$$= b \int\limits_{\zeta=a}^{\zeta=a+t_f} \zeta d\zeta + t_w \int\limits_{\zeta=z}^{\zeta=a} \zeta d\zeta \qquad (2.142)$$

$$= bct_f + (a^2 - z^2)\frac{t_w}{2}. \qquad (2.143)$$

Fig. 2.38 Schematic representation of the shear stress distribution in the I-profile

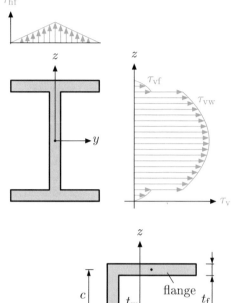

Fig. 2.39 C-profile for the calculation of the shear stress distribution

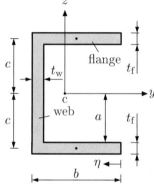

Thus, we obtain the final expression for the vertical shear stress in the web of a thin-walled I-profile as follows:

$$\tau_{vw}(z) = \frac{Q_y(x)}{I_y}\left(bc\frac{t_f}{t_w} + \frac{a^2}{2}\left(1 - \frac{z^2}{a^2}\right)\right). \tag{2.144}$$

The entire shear stress distribution based on Eqs. (2.136) and (2.144) is shown in Fig. 2.38. There is in addition for completeness a vertical shear force (τ_{vf}) in the flange indicated. However, this component is normally disregarded for thin-walled sections.

Let us focus in the following on another thin-walled beam, i.e. the C-profile shown in Fig. 2.39.

The calculation of the shear stress distributions in the flanges and the web can be performed under the same assumptions and procedure as for the I-profile and one obtains after a short calculation the following expressions:

$$\tau_{\mathrm{hf}}(\eta) = \frac{Q_y(x)c}{I_y} \times \eta \quad (0 \le \eta \le b) \tag{2.145}$$

$$\tau_{\mathrm{vw}}(z) = \frac{Q_y(x)}{I_y} \left(bc \frac{t_{\mathrm{f}}}{t_{\mathrm{w}}} + \frac{a^2}{2} \left(1 - \frac{z^2}{a^2} \right) \right) \quad (-a \le z \le a). \tag{2.146}$$

The evaluation of Eqs. (2.145) and (2.146) requires the axial second moment of area I_y. The calculation of this quantity requires the coordinates of the centroid ('c'). Since this C-profile is singly symmetric, it remains to calculate the horizontal position of the centroid. As a first calculation step, let us split the C-profile into three simple shapes, i.e., three rectangles, as shown in Fig. 2.40. In addition, an initial y'-z' coordinate system is introduced with its origin on the symmetry line at the left-hand boundary.

The coordinates of the single centroids as wells as the corresponding surface areas are summarized in Table 2.13.

The coordinate y'_c of the centroid can be obtained from Eq. (B.4) as:

$$y'_c = \frac{y'_1 A_1 + y'_2 A_2 + y'_3 A_3}{A_1 + A_2 + A_3}, \tag{2.147}$$

$$= \frac{b^2 t_{\mathrm{f}} + a t_{\mathrm{w}}^2}{2(b t_{\mathrm{f}} + a t_{\mathrm{w}})}. \tag{2.148}$$

Fig. 2.40 Subdivision of the C-profile into three subsections and initial y'-z' coordinate system for the determination of the centroid

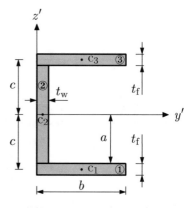

Table 2.13 Position of centroids in the y'-z' coordinate system and corresponding surface areas

Surface i	y'_i	A_i
1	$\dfrac{b}{2}$	$b t_{\mathrm{f}}$
2	$\dfrac{t_{\mathrm{w}}}{2}$	$2 a t_{\mathrm{w}}$
3	$\dfrac{b}{2}$	$b t_{\mathrm{f}}$

Fig. 2.41 Schematic representation of the shear stress distribution in the C-profile

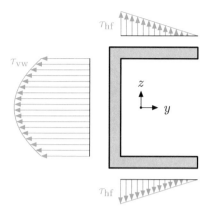

The next step is to calculate the second moment of area in the y-z coordinate system (see Fig. 2.39). Based on the second moments for simple shapes provided in Table 2.6 and the parallel axis theorem provided in Sect. B.3, the following relationship is obtained:

$$I_y = I_{y,1} + I_{y,2} + I_{y,3} \tag{2.149}$$

$$= \frac{1}{12} t_w (2a)^3 + 2 \times \frac{1}{12} b(t_f)^3 + 2 \times (a + t_f/2)^2 b t_f$$

$$= \frac{2}{3} a^3 t_w + \frac{2}{3} b t_f^3 + 2a^2 b t_f + 2ab t_f^2 . \tag{2.150}$$

The entire shear stress distribution based on Eqs. (2.145) and (2.146) is shown in Fig. 2.41.

Let us now further simplify the considered case of the C-profile and assume that $t_f \ll b$ and $t_w \ll c$. Furthermore, let us assume that $b = c$. This allows to use the following statements: $c \approx a$, $b \approx a$, and $t \approx t_f \approx t_w$. Thus, the coordinate of the centroid according to Eq. (2.148) can be simplified for this special case to the following expression:

$$y_c' \approx \frac{a}{4}, \tag{2.151}$$

as well as the expression for the second moment of area according to Eq. (2.150):

$$I_y \approx \frac{8}{3} a^3 t . \tag{2.152}$$

Let us now assume in addition a cantilever beam configuration as shown in Fig. 2.42a. It is assumed that the line of action of the vertical force F_z passes through the centroid 'c'. The flow of the shear stress in the thin section is schematically indicated in Fig. 2.42b. It is easy to understand that the resultants of the shear stress distributions

(a)

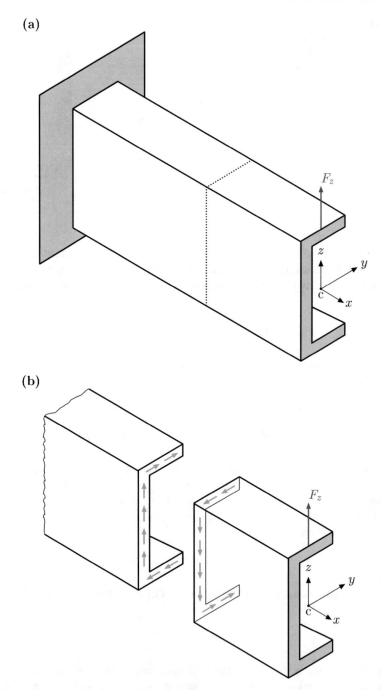

(b)

Fig. 2.42 Cantilever beam with C-profile: **a** general configuration with external force F_z and **b** part for the determination of the shear flow direction

Fig. 2.43 Thin-walled
C-profile: determination of
the shear center 'sc'

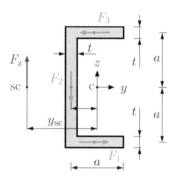

produce a twisting moment with respect to the centroid. The horizontal forces from
the shear stresses in the flanges (consider the opposite direction in the upper and
lower web but with the same magnitude) as well as the resultant from the shear stress
in the web have a lever with respect to the centroid and they would result in a torsional
deformation around the x-axis. Only the external force will have *no* moment action
with respect to the centroid as long as its line of action passes through the centroid.

To avoid this twisting of the beam, the line of action of the external force F_z must
pass through another point, i.e. the so-called shear center 'sc', see Fig. 2.43.

It holds for the horizontal resultants that $F_3 = -F_1$ and their magnitudes (i.e., simple consideration of the area) can be obtained from Eq. (2.145) under consideration
of Eq. (2.152) as:

$$F_1 = -F_3 = \frac{1}{2}\tau_{hf}(\eta = a)at = \frac{3}{16}Q_z. \tag{2.153}$$

The calculation of the resultant F_2 requires the following integration:

$$F_2 = \int_{-a}^{a}\int_{-t/2}^{t/2} \tau_{vw}(z)\,dA = \int_{-a}^{a}\int_{-t/2}^{t/2}\left(\frac{3Q_z}{16at}\left(3 - \frac{z^2}{a^2}\right)\right)dA = Q_z. \tag{2.154}$$

Thus, we can finally state the moment equilibrium with respect to the centroid in
order to calculate the vertical position of the shear center:

$$\sum_{i=1}^{3}M_i = 0 \quad \Leftrightarrow \quad -F_z|y_{sc}| + F_2\frac{a}{4} + F_1a + F_1a = 0, \tag{2.155}$$

or rearranged for the location of the shear center:

$$|y_{sc}| = \frac{F_2\dfrac{a}{4} + F_1a + F_1a}{F_z} = \frac{5}{8}a. \tag{2.156}$$

(a)   (b)   (c)

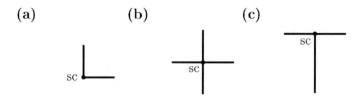

Fig. 2.44 Location of the shear center for some common cross sections: **a** L-, **b** +- and **c** T-profile

Further locations of the shear center for some common cross sections are collected in Fig. 2.44 and the following general statements in regards to the location of the shear center can be provided:

- In symmetric cross section, the shear center lies in the plane of symmetry.
- In doubly symmetric cross section, the shear center lies in the centroid.

2.8   Supplementary Problems

2.1   Cantilever beam with different end loads and deformations
Calculate the analytical solution for the deflection $u_z(x)$ and rotation $\varphi_y(x)$ of the cantilever beams shown in Fig. 2.45. Start your derivation from the fourth-order differential equation. It can be assumed for this exercise that the bending stiffness EI_y is constant. Calculate in addition for all four cases the reactions at the fixed support and the distributions of the bending moment and shear force.

2.2   Beam fixed at both ends: analytical solution of the deformations
Calculate the analytical solution for the deflection $u_z(x)$ and slope $\varphi_y(x)$ of the Euler–Bernoulli beams shown in Fig. 2.46 based on the fourth-order differential equation

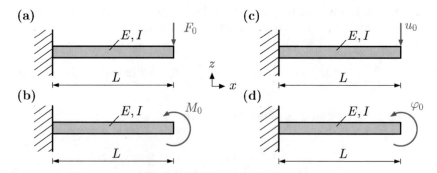

Fig. 2.45 Cantilever beam with different end loads and deformations: **a** single force; **b** single moment; **c** displacement; **d** rotation

(a)   **(b)**

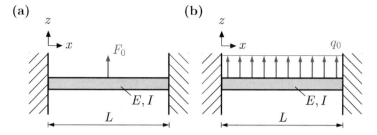

Fig. 2.46 Euler–Bernoulli beam fixed at both ends: **a** single force case; **b** distributed load case

Fig. 2.47 Euler–Bernoulli
beam with a quadratic
distributed load

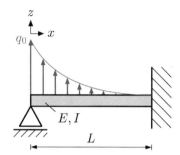

given in Table 2.7. Determine in addition the maximum deflection and slope. It can
be assumed for this exercise that the bending stiffness EI_y is constant.

2.3 Euler–Bernoulli beam with quadratic distributed load: analytical solution of the bending line

Given is the Euler–Bernoulli beam shown in Fig. 2.47. The length of the beam is equal
to L and the bending stiffness EI_y is constant. The beam is loaded by a quadratic
distributed load (maximum value q_0):

$$q(x) = q_0 \left(\frac{x}{L} - 1 \right)^2 . \tag{2.157}$$

Calculate the analytical solution for the deflection $u_z(x)$ and the internal bending
moment distribution $M_y(x)$ starting from the fourth-order differential equation given
in Table 2.7. In addition, provide schematic sketches of both distributions. Shear
effects can be disregarded for this problem.

2.4 Unsymmetrical bending of a Z-profile

A cantilever beam with a Z-shaped cross section as shown in Fig. 2.48 is loaded
by a single force F_0 at the free tip. The line of action of this force passes through
the centroid of the cross section and results in a bending moment of magnitude of

Fig. 2.48 Z-profile for unsymmetrical bending problem

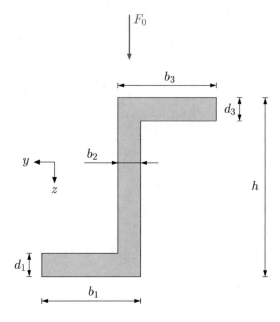

8.4×10^6 Nmm. Shear stresses due to shear forces, the dead weight as well as stress concentrations at the support can be disregarded.

The geometrical dimensions of the problem are: $b_1 = 90$ mm, $d_1 = 20$ mm, $b_2 = 20$ mm, $b_3 = 90$ mm, $d_3 = 20$ mm, $h = 150$ mm.

Calculate the position and the magnitude of the maximum tensile and compressive stress.

2.5 Calculation of the shear stress distribution in a circular cross section

Given is a beam with circular cross section of radius R. Calculate the distribution of the shear stress $\tau_{xz}(z)$ over the cross section under the influence of a shear force $Q_z(x)$. Consider the distribution in the middle of the section, i.e. for $y = 0$.

2.6 Calculation of the shear stress distribution in a triangular cross section

Given is a beam with a triangular cross section, i.e., an equilateral triangle of side length a, see Fig. 2.49.

Calculate the distribution of the shear stress $\tau_{xz}(z)$ over the cross section under the influence of a shear force $Q_z(x)$. Consider the distribution in the middle of the section, i.e. for $y = 0$.

2.7 Calculation of the shear stress distributions in circular ring cross sections

Given are beams with circular ring cross sections, see Fig. 2.50.

Calculate the distribution of the shear stress $\tau_{xz}(\varphi)$ in the cross sections under the influence of a shear force $Q_z(x)$. Consider that the cross sections of thickness t and

Fig. 2.49 Triangular cross section of a beam

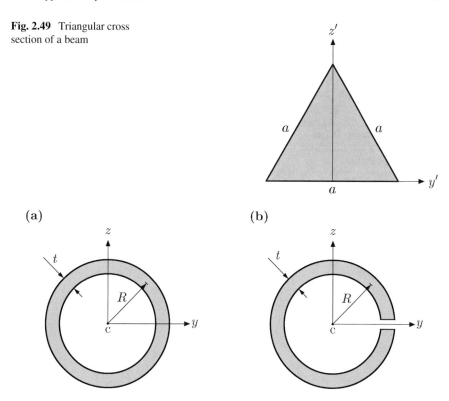

(a)   (b)

Fig. 2.50 Thin-walled circular ring cross sections of breams: **a** continuous and **b** slotted

average radius R are considered as thin. Furthermore, determine the location of the shear center for both configurations. Consider a cantilever beam configuration with a tip load of F_z in positive direction of the z-axis.

References

1. Altenbach H, Altenbach J, Naumenko K (2016) Ebene Flächentragwerke: Grundlagen der Modellierung und Berechnung von Scheiben und Platten. Springer, Berlin
2. Altenbach H, Öchsner A (eds) (2020) Encyclopedia of continuum mechanics. Springer, Berlin
3. Budynas RG (1999) Advanced strength and applied stress analysis. McGraw-Hill Book, Singapore
4. Gould PL (1988) Analysis of shells and plates. Springer, New York
5. Gross D, Hauger W, Schröder J, Wall WA (2009) Technische Mechanik 2: Elastostatik. Springer, Berlin
6. Hartmann F, Katz C (2007) Structural analysis with finite elements. Springer, Berlin
7. Heyman J (1998) Structural analysis: a historical approach. Cambridge University Press, Cambridge

8. Hibbeler RC (2008) Mechanics of materials. Prentice Hall, Singapore
9. Öchsner A (2014) Elasto-plasticity of frame structure elements: modeling and simulation of rods and beams. Springer, Berlin
10. Öchsner A, Merkel M (2018) One-dimensional finite elements: an introduction to the FE method. Springer, Cham
11. Öchsner A (2020) Computational statics and dynamics: an introduction based on the finite element method. Springer, Singapore
12. Reddy JN (2004) Mechanics of laminated composite plates and shells: theory and analysis. CRC Press, Boca Raton
13. Szabó I (2003) Einführung in die Technische Mechanik: Nach Vorlesungen István Szabó. Springer, Berlin
14. Timoshenko S, Woinowsky-Krieger S (1959) Theory of plates and shells. McGraw-Hill Book Company, New York
15. Winkler E (1867) Die Lehre von der Elasticität und Festigkeit mit besonderer Rücksicht auf ihre Anwendung in der Technik. H. Dominicus, Prag

Chapter 3
Timoshenko Beam Theory

Abstract This chapter presents the analytical description of thick, or so-called shear-flexible, beam members according to the Timoshenko theory. Based on the three basic equations of continuum mechanics, i.e., the kinematics relationship, the constitutive law, and the equilibrium equation, the partial differential equations, which describe the physical problem, are presented. All equations are introduced for single plane bending in the x-y plane as well as the x-z plane. Analytical solutions of the partial differential equations are given for simple cases. In addition, this chapter treats the case of unsymmetrical bending.

3.1 Introduction

The general difference regarding the deformation of a beam with and without shear influence has already been discussed in Sect. 2.1. In this section, the shear influence on the deformation is considered with the help of the Timoshenko beam theory [14, 15]. Within the framework of the following remarks, the definition of the shear strain and the relation between shear force and shear stress will first be covered.

3.2 Deformation in the x-y Plane

For the derivation of the equation for the shear strain in the x-y plane, the infinitesimal rectangular beam element $ABCD$, shown in Fig. 3.1, is considered, which deforms under the influence of a pure shear stress. Here, a change of the angle of the original right angles as well as a change in the lengths of the edges occurs.

The deformation of the point A can be described based on the displacement fields $u_x(x, y)$ and $u_y(x, y)$. These two functions of *two* variables can be expanded

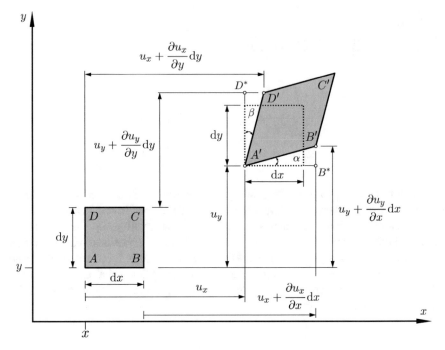

Fig. 3.1 Definition of the shear strain γ_{xy} in the x-y plane at an infinitesimal beam element

into Taylor's series[1] of first-order around point A to approximately calculate the deformations of the points B and D:

$$u_{x,B} = u_x(x + \mathrm{d}x, y) = u_x(x, y) + \frac{\partial u_x(x, y)}{\partial x}\mathrm{d}x + \frac{\partial u_x(x, y)}{\partial y}\mathrm{d}y, \qquad (3.1)$$

$$u_{y,B} = u_y(x + \mathrm{d}x, y) = u_y(x, y) + \frac{\partial u_y(x, y)}{\partial x}\mathrm{d}x + \frac{\partial u_y(x, y)}{\partial y}\mathrm{d}y, \qquad (3.2)$$

or alternatively

$$u_{x,D} = u_x(x, y + \mathrm{d}y) = u_x(x, y) + \frac{\partial u_x(x, y)}{\partial x}\mathrm{d}x + \frac{\partial u_x(x, y)}{\partial y}\mathrm{d}y, \qquad (3.3)$$

$$u_{y,D} = u_y(x, y + \mathrm{d}y) = u_y(x, y) + \frac{\partial u_y(x, y)}{\partial x}\mathrm{d}x + \frac{\partial u_y(x, y)}{\partial y}\mathrm{d}y. \qquad (3.4)$$

---

[1]For a function $f(x, y)$ of two variables usually a Taylor's series expansion of first-order is formulated around the point (x_0, y_0) as follows: $f(x, y) = f(x_0 + \mathrm{d}x, y_0 + \mathrm{d}y) \approx f(x_0, y_0) + \left(\frac{\partial f}{\partial x}\right)_{x_0, y_0} \times (x - x_0) + \left(\frac{\partial f}{\partial y}\right)_{x_0, y_0} \times (y - y_0)$.

In Eqs. (3.1) up to (3.4), $u_x(x, y)$ and $u_y(x, y)$ represent the so-called rigid-body displacements, which do not cause a deformation. If one considers that point B has the coordinates $(x + dx, y)$ and D the coordinates $(x, y + dy)$, the following results:

$$u_{x,B} = u_x(x, y) + \frac{\partial u_x(x, y)}{\partial x} dx , \qquad (3.5)$$

$$u_{y,B} = u_y(x, y) + \frac{\partial u_y(x, y)}{\partial x} dx , \qquad (3.6)$$

or alternatively

$$u_{x,D} = u_x(x, y) + \frac{\partial u_x(x, y)}{\partial y} dy , \qquad (3.7)$$

$$u_{y,D} = u_y(x, y) + \frac{\partial u_y(x, y)}{\partial y} dy . \qquad (3.8)$$

The total shear strain γ_{xy} of the deformed beam element $A'B'C'D'$ results, according to Fig. 3.1, from the sum of the angles α and β. The two angles can be identified in the rectangle, which is deformed to a rhombus. Under consideration of the two right-angled triangles $A'D^*D'$ and $A'B^*B'$, these two angles can be expressed as:

$$\tan \alpha = \frac{\dfrac{\partial u_y(x, y)}{\partial x} dx}{dx + \dfrac{\partial u_x(x, y)}{\partial x} dx} \quad \text{and} \quad \tan \beta = \frac{\dfrac{\partial u_x(x, y)}{\partial y} dy}{dy + \dfrac{\partial u_y(x, y)}{\partial y} dy} . \qquad (3.9)$$

It holds approximately for small deformations that $\tan \alpha \approx \alpha$ and $\tan \beta \approx \beta$ or alternatively $\frac{\partial u_x}{\partial x} \ll 1$ and $\frac{\partial u_y}{\partial y} \ll 1$, so that the following expression results for the shear strain in the x-y plane:

$$\gamma_{xy} = \alpha + \beta \approx \frac{\partial u_y(x, y)}{\partial x} + \frac{\partial u_x(x, y)}{\partial y} . \qquad (3.10)$$

This total change of the angle is also called the engineering shear strain. In contrast to this, the expression $\varepsilon_{xy} = \frac{1}{2} \gamma_{xy} = \frac{1}{2}(\frac{\partial u_y}{\partial x} + \frac{\partial u_x}{\partial y})$ is known as the tensorial definition (tensor shear strain) in the literature [18]. Due to the symmetry of the strain tensor, the identity $\gamma_{ij} = \gamma_{ji}$ applies to the tensor elements outside the main diagonal.

The algebraic sign of the shear strain needs to be explained in the following with the help of Fig. 3.2 for the special case that only one shear force acts in parallel to the y-axis. If a shear force acts in the direction of the positive y-axis at the right-hand face—hence a positive shear force distribution is being assumed at this point—,

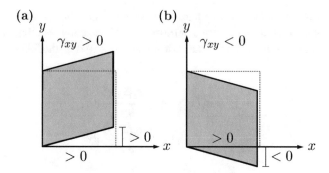

Fig. 3.2 Definition of a **a** positive and **b** negative shear strain in the x-y plane

according to Fig. 3.2a under consideration of Eq. (3.10) a positive shear strain results. In a similar way, a negative shear force distribution leads to a negative shear strain according to Fig. 3.2b.

It has already been mentioned in Sect. 2.1 that the shear stress distribution is variable over the cross-section. As an example, the parabolic shear stress distribution was illustrated over a rectangular cross section in Fig. 2.2. Based on Hooke's law for a one-dimensional shear stress state, it can be derived that the shear strain has to exhibit a corresponding parabolic course. From the shear stress distribution in the cross-sectional area at location x of the beam,[2] one receives the acting shear force through integration over the cross section as:

$$Q_y(x) = \int_A \tau_{xy}(x, y, z)\, \mathrm{d}A .$$ (3.11)

However, to simplify the problem, it is assumed for the Timoshenko beam that an equivalent *constant* shear stress and strain act at location x, see Fig. 3.11:

$$\tau_{xy}(x, y, z) \rightarrow \tau_{xy}(x) .$$ (3.12)

This constant shear stress results from the shear force, which acts in an equivalent cross-sectional area, the so-called shear area A_s (Fig. 3.3):

$$\tau_{xy}(x) = \frac{Q_y(x)}{A_s} ,$$ (3.13)

---

[2]A closer analysis of the shear stress distribution in the cross-sectional area shows that the shear stress does not just alter over the height of the beam but also through the width of the beam. If the width of the beam is small when compared to the height, only a small change along the width occurs and one can assume in the first approximation a constant shear stress throughout the width: $\tau_{xy}(y, z) \rightarrow \tau_{xy}(y)$. See for example [3, 17].

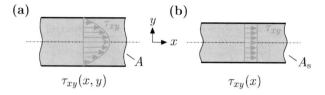

Fig. 3.3 Shear stress distribution in the x-y plane: **a** real distribution for a rectangular cross section and **b** Timoshenko's approximation

Table 3.1 Comparison of shear correction factor values for a rectangular cross section based on different approaches

k_s	Comment	Reference
$\frac{2}{3}$	–	[14, 16]
$0.833\left(=\frac{5}{6}\right)$	$\nu = 0.0$	[4]
0.850	$\nu = 0.3$	
0.870	$\nu = 0.5$	

whereupon the relation between the shear area A_s and the actual cross-sectional area A_s is referred to as the shear correction factor k_s:

$$k_s = \frac{A_s}{A}. \tag{3.14}$$

Different assumptions can be made to calculate the shear correction factor [4]. As an example, it can be demanded [2] that the elastic strain energy of the equivalent shear stress has to be identical with the energy, which results from the acting shear stress distribution in the actual cross-sectional-area. A comparison for a rectangular cross section is presented in Table 3.1.

Different geometric characteristics of simple geometric cross-sections—including the shear correction factor[3]—are collected in Table 3.2 [5, 19]. Further details regarding the shear correction factor for arbitrary cross-sections can be taken from [6].

It is obvious that the equivalent constant shear stress can alter along the center line of the beam, in case the shear force along the center line of the beam changes. The attribute 'constant' thus just refers to the cross-sectional area at location x and the equivalent constant shear stress is therefore in general a function of the coordinate of length for the Timoshenko beam:

$$\tau_{xy} = \tau_{xy}(x). \tag{3.15}$$

---

[3]It should be noted that the so-called form factor for shear is also known in the literature. This results as the reciprocal of the shear correction factor.

Table 3.2 Characteristics of different cross sections in the y-y plane. I_z: axial second moment of area; A: cross-sectional area; k_s: shear correction factor. Adapted from [19]

Cross-section	I_z	A	k_s
circle, $D = 2R$	$\dfrac{\pi R^4}{4}$	πR^2	$\dfrac{9}{10}$
hollow circle, $D = 2R$, t	$\pi R^3 t$	$2\pi R t$	0.5
rectangle, h, b	$\dfrac{hb^3}{12}$	hb	$\dfrac{5}{6}$
box, t_f, t_w, h, b	$\dfrac{b^2}{6}(bt_f + 3ht_w)$	$2(bt_f + ht_w)$	$\dfrac{2ht_w}{A}$
I-section, t_f, t_w, h, b	$\dfrac{b^3 t_f}{6}$	$ht_w + 2bt_f$	$\dfrac{ht_w}{A}$

The so-called Timoshenko beam can be generated by superposing a shear deformation on an Euler–Bernoulli beam according to Fig. 3.4.

One can see that Bernoulli's hypothesis is partly no longer fulfilled for the Timoshenko beam: Plane cross sections remain plane after the deformation. However, a cross section which stood at right angles on the beam axis before the deformation is not at right angles on the beam axis after the deformation. If the demand for planeness of the cross sections is also given up, one reaches theories of higher-order [7, 11, 12], at which, for example, a parabolic course of the shear strain and stress in the displacement field are considered, see Fig. 3.5. Therefore, a shear correction factor is not required for these theories of higher-order.

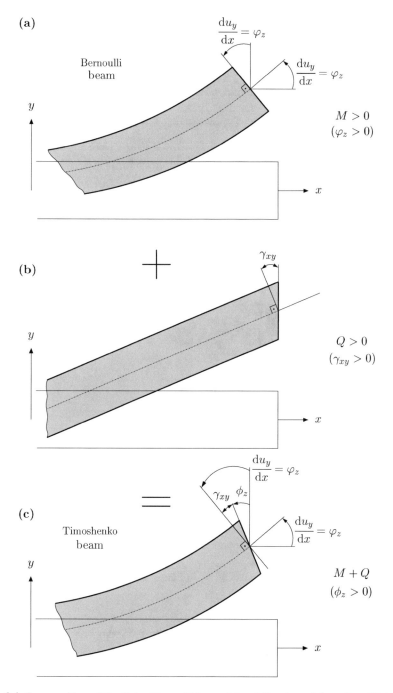

Fig. 3.4 Superposition of the Euler–Bernoulli beam (**a**) and the shear deformation (**b**) to the Timoshenko beam (**c**) in the *x-y* plane. See Fig. 3.2 for clarification on the sign of γ_{xy}. Note that the deformation is exaggerated for better illustration

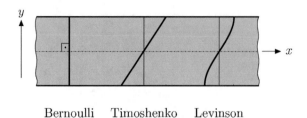

Fig. 3.5 Deformation of originally plane cross sections in the x-y plane for the Euler–Bernoulli beam (left), the Timoshenko beam (middle), and the Levinson beam (right) [13]

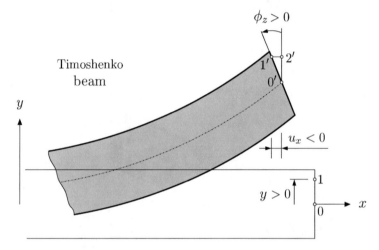

Fig. 3.6 Derivation of the kinematics relation in the x-y plane. Note that the deformation is exaggerated for better illustration

3.2.1 Kinematics

According to the alternative derivation in Sect. 2.2.1, the kinematics relation can also be derived for the beam with shear action, by considering the angle ϕ_z instead of the angle φ_z, see Figs. 3.4c and 3.6.

Following an equivalent procedure as in Sect. 2.2.1, the corresponding relationships are obtained:

$$\sin \phi_z = \frac{-u_x}{y} \approx \phi_z \quad \text{or} \quad u_x = -y\phi_z\,, \tag{3.16}$$

wherefrom, via the general relation for the strain, meaning $\varepsilon_x = \mathrm{d}u_x/\mathrm{d}x$, the kinematics relation results through differentiation with respect to the x-coordinate:

$$\varepsilon_x = -y \frac{\mathrm{d}\phi_z}{\mathrm{d}x}. \tag{3.17}$$

Note that $\phi_z \rightarrow \varphi_z = \frac{\mathrm{d}u_y}{\mathrm{d}x}$ results from neglecting the shear deformation and a relation according to Eq. (2.16) results as a special case. Furthermore, the following relation between the angles can be derived from Fig. 3.4c

$$\phi_z = \varphi_z - \gamma_{xy} = \frac{\mathrm{d}u_y}{\mathrm{d}x} - \gamma_{xy}, \tag{3.18}$$

which complements the set of the kinematics relations. It needs to be remarked that at this point the so-called bending line was considered. Therefore, the displacement field u_y is only a function of *one* variable: $u_y = u_y(x)$.

3.2.2 Constitutive Equation

For the consideration of the constitutive relation, Hooke's law for a one-dimensional normal stress state and for a one-dimensional shear stress state is used:

$$\sigma_x = E\varepsilon_x, \tag{3.19}$$

$$\tau_{xy} = G\gamma_{xy}, \tag{3.20}$$

whereupon the shear modulus G can be calculated for isotropic materials based on the Young's modulus E and the Poisson's ratio ν as:

$$G = \frac{E}{2(1+\nu)}. \tag{3.21}$$

According to the equilibrium configuration of Fig. 2.9 and Eq. (2.22), the relation between the internal moment and the bending stress can be used for the Timoshenko beam as follows:

$$\mathrm{d}M_z = (-y)(+\sigma_x)\mathrm{d}A, \tag{3.22}$$

or alternatively after integration under the consideration of the constitutive equation (3.19) and the kinematics relation (3.17):

$$M_z(x) = +EI_z \frac{\mathrm{d}\phi_z(x)}{\mathrm{d}x}. \tag{3.23}$$

The relation between shear force and cross-sectional rotation results from the equilibrium equation (3.28) as:

$$Q_y(x) = -\frac{dM_z(x)}{dx} - m_z(x) = -EI_z \frac{d^2\phi_z(x)}{dx^2} - m_z(x). \qquad (3.24)$$

Alternatively, the internal shear force results from Eq. (3.13) and the application of the constitutive relation (3.20) as:

$$Q_y(x) = \tau_{xy}A_s = k_sAG\gamma_{xy}(x), \qquad (3.25)$$

or under consideration of the kinematics relation (3.18) as a solely function of the deformations $u_y(x)$ and $\phi_z(x)$:

$$Q_y(x) = k_sAG\left(\frac{du_y(x)}{dx} - \phi_z(x)\right). \qquad (3.26)$$

Similar to Sect. 2.2.2, one can use the stress resultants to formulate the constitutive law based on the generalized stresses $s = \begin{bmatrix} M_z, Q_y \end{bmatrix}^T$ and generalized strains $e = \begin{bmatrix} \frac{d\phi_z}{dx}, \frac{du_y}{dx} - \phi_z \end{bmatrix}^T$, see Fig. 3.7. The generalized quantities reveal the advantage that they are not a function of the vertical coordinate.

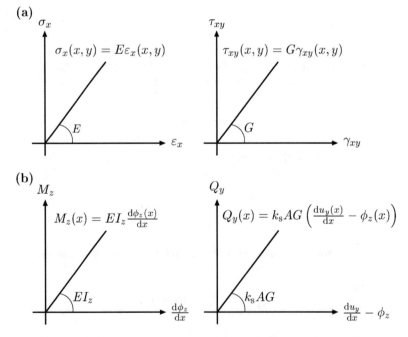

Fig. 3.7 Formulation of the constitutive law based on **a** classical stress-strain and **b** generalized-stress-generalized-strain relations (x-y plane)

3.2.3 Equilibrium

The derivation of the equilibrium condition for the Timoshenko beam is identical with the derivation for the Euler–Bernoulli beam according to Sect. 2.2.3:

$$\frac{dQ_y(x)}{dx} = -q_y(x),$$ (3.27)

$$\frac{dM_z(x)}{dx} = -Q_y(x) - m_z(x).$$ (3.28)

Before looking in more detail at the differential equations of the bending line, let us summarize the basic equations for the Timoshenko beam in Table 3.3. Note that the normal stress and normal strain are functions of both spatial coordinates, i.e. x and y. However, the shear stress and shear strain are only dependent on the x-coordinate, since an equivalent *constant* shear stress has been introduced over the cross section as an approximation of the Timoshenko beam theory.

3.2.4 Differential Equation

Within the previous section, the relation between the internal moment and the cross-sectional rotation was derived from the normal stress distribution with the help of Hooke's law, see Eq. (3.23). Differentiation of this relation with respect to the x-coordinate leads to the following expression

$$\frac{dM_z(x)}{dx} = \frac{d}{dx}\left(EI_z\frac{d\phi_z(x)}{dx}\right),$$ (3.29)

which can be transformed with the help of the equilibrium relation (3.28), the constitutive equation (3.20), and the relation for the shear stress according to (3.13) and (3.14) to

Table 3.3 Elementary basic equations for the bending of a Timoshenko beam in the $x - y$ plane (x-axis: right facing; y-axis: upward facing)

Relation	Equation
Kinematics	$\varepsilon_x(x, y) = -y\dfrac{d\phi_z(x)}{dx}$ and $\phi_z(x) = \dfrac{du_y(x)}{dx} - \gamma_{xy}(x)$
Equilibrium	$\dfrac{dQ_y(x)}{dx} = -q_y(x); \quad \dfrac{dM_z(x)}{dx} = -Q_y(x) - m_z(x)$
Constitution	$\sigma_x(x, y) = E\varepsilon_x(x, y)$ and $\tau_{xy}(x) = G\gamma_{xy}(x)$

$$\frac{\mathrm{d}}{\mathrm{d}x}\left(EI_z \frac{\mathrm{d}\phi_z(x)}{\mathrm{d}x}\right) = -k_s GA\gamma_{xz}(x) - m_z(x). \tag{3.30}$$

If the kinematics relation (3.18) is considered in the last equation, the so-called bending differential equation results in:

$$\frac{\mathrm{d}}{\mathrm{d}x}\left(EI_z \frac{\mathrm{d}\phi_z(x)}{\mathrm{d}x}\right) + k_s GA\left(\frac{\mathrm{d}u_y(x)}{\mathrm{d}x} - \phi_z(x)\right) = -m_z(x). \tag{3.31}$$

Considering the shear stress according to Eqs. (3.13) and (3.14) in the expression of Hooke's law according to (3.20), one obtains

$$Q_y(x) = k_s AG\gamma_{xy}(x). \tag{3.32}$$

Introducing the equilibrium relation (3.28) and the kinematics relation (3.18) in the last equation gives:

$$-\frac{\mathrm{d}M_z(x)}{\mathrm{d}x} - m_z(x) = +k_s AG\left(\frac{\mathrm{d}u_y(x)}{\mathrm{d}x} - \phi_z(x)\right). \tag{3.33}$$

After differentiation and the consideration of the equilibrium relations according to Eqs. (3.27) and (3.28), the so-called shear differential equation results finally in:

$$\frac{\mathrm{d}}{\mathrm{d}x}\left[k_s AG\left(\frac{\mathrm{d}u_y(x)}{\mathrm{d}x} - \phi_z(x)\right)\right] = -q_y(x). \tag{3.34}$$

Therefore, the shear flexible Timoshenko beam is described through the following two coupled differential equations of second order:

$$\frac{\mathrm{d}}{\mathrm{d}x}\left(EI_z \frac{\mathrm{d}\phi_z(x)}{\mathrm{d}x}\right) + k_s AG\left(\frac{\mathrm{d}u_y(x)}{\mathrm{d}x} - \phi_z(x)\right) = -m_z(x), \tag{3.35}$$

$$\frac{\mathrm{d}}{\mathrm{d}x}\left[k_s AG\left(\frac{\mathrm{d}u_y(x)}{\mathrm{d}x} - \phi_z(x)\right)\right] = -q_y(x). \tag{3.36}$$

This system contains two unknown functions, namely the deflection $u_y(x)$ and the cross-sectional rotation $\phi_z(x)$. Boundary conditions must be formulated for both functions to be able to solve the system of differential equations for a specific problem.

Different formulations of these coupled differential equations are collected in Table 3.4 where different types of loadings, geometry and bedding are differentiated. The last case in Table 3.4 refers again to the elastic or Winkler foundation of a beam, [20]. The elastic foundation modulus k has in the case of beams the unit of force per unit area.

Table 3.4 Different formulations of the partial differential equation for a Timoshenko beam in the x-y plane (x-axis: right facing; y-axis: upward facing)

Configuration	Partial differential equation
E, I_z, A, G, k_s	$EI_z\dfrac{d^2\phi_z}{dx^2} + k_sGA\left(\dfrac{du_y}{dx} - \phi_z\right) = 0$ $k_sGA\left(\dfrac{d^2u_y}{dx^2} - \dfrac{d\phi_z}{dx}\right) = 0$
$E(x), I_z(x)$ $k_s(x), A(x), G(x)$	$\dfrac{d}{dx}\left(E(x)I_z(x)\dfrac{d\phi_z}{dx}\right) + k_s(x)G(x)A(x)\left(\dfrac{du_y}{dx} - \phi_z\right) = 0$ $\dfrac{d}{dx}\left[k_s(x)G(x)A(x)\left(\dfrac{du_y}{dx} - \phi_z\right)\right] = 0$
$q_y(x)$	$EI_z\dfrac{d^2\phi_z}{dx^2} + k_sGA\left(\dfrac{du_y}{dx} - \phi_z\right) = 0$ $k_sGA\left(\dfrac{d^2u_y}{dx^2} - \dfrac{d\phi_z}{dx}\right) = -q_y(x)$
$m_z(x)$	$EI_z\dfrac{d^2\phi_z}{dx^2} + k_sGA\left(\dfrac{du_y}{dx} - \phi_z\right) = -m_z(x)$ $k_sGA\left(\dfrac{d^2u_y}{dx^2}\dfrac{d\phi_z}{dx}\right) = 0$
$k(x)$	$EI_z\dfrac{d^2\phi_z}{dx^2} + k_sGA\left(\dfrac{du_y}{dx} - \phi_z\right) = 0$ $k_sGA\left(\dfrac{d^2u_y}{dx^2} - \dfrac{d\phi_z}{dx}\right) = k(x)u_y$

A single-equation description for the Timoshenko beam can be obtained under the assumption of constant material (E, G) and geometrical (I_z, A, k_s) properties: Rearranging and two-times differentiation of Eq. (3.36) gives:

$$\frac{d\phi_z(x)}{dx} = +\frac{d^2u_y(x)}{dx^2} + \frac{q_y(x)}{k_sGA}, \tag{3.37}$$

$$\frac{d^3\phi_z(x)}{dx^3} = +\frac{d^4u_y(x)}{dx^4} + \frac{d^2q_y(x)}{k_sGAdx^2}. \tag{3.38}$$

One-time differentiation of Eq. (3.35) gives:

$$E I_z \frac{d^3 \phi_z(x)}{dx^3} + k_s A G \left(\frac{d^2 u_y(x)}{dx^2} - \frac{d\phi_z(x)}{dx} \right) = -\frac{dm_z(x)}{dx}. \tag{3.39}$$

Inserting Eq. (3.37) into (3.39) and consideration of (3.38) gives finally the following expression:

$$E I_z \frac{d^4 u_y(x)}{dx^4} = q_y(x) - \frac{dm_z(x)}{dx} - \frac{E I_z}{k_s A G} \frac{d^2 q_y(x)}{dx^2}. \tag{3.40}$$

The last equation reduces for shear-rigid beams, i.e. $k_s A G \to \infty$, to the classical Euler–Bernoulli formulation as given in Table 2.4.

If we replace in the previous formulations the first-order derivative, i.e. $\frac{d\cdots}{dx}$, by a formal operator symbol, i.e. the $\mathcal{L}_1$–matrix, then the basic equations of the Timoshenko beam can be stated in a more formal way as given in Table 3.5. Such a matrix formulation is suitable for the derivation of the principal finite element equation based on the weighted residual method [9, 10].

Table 3.5 Different formulations of the basic equations for a Timoshenko beam (bending in the x-y plane; x-axis along the principal beam axis). E: Young's modulus; G: shear modulus; A: cross-sectional area; I_y: second moment of area; k_s: shear correction factor; q_y: length-specific distributed force; m_y: length-specific distributed moment; e: generalized strains; s^*: generalized stresses

Specific formulation	General formulation [1]
Kinematics	
$\begin{bmatrix} \frac{du_y}{dx} - \phi_z \\ \frac{d\phi_x}{dx} \end{bmatrix} = \begin{bmatrix} \frac{d}{dx} & -1 \\ 0 & \frac{d}{dx} \end{bmatrix} \begin{bmatrix} u_y \\ \phi_z \end{bmatrix}$	$e = \mathcal{L}_1 * u$
Constitution	
$\begin{bmatrix} Q_y \\ M_z \end{bmatrix} = \begin{bmatrix} k_s A G & 0 \\ 0 & E I_z \end{bmatrix} \begin{bmatrix} \frac{du_y}{dx} - \phi_z \\ \frac{d\phi_z}{dx} \end{bmatrix}$	$s^* = D^* e$
Equilibrium	
$\begin{bmatrix} \frac{d}{dx} & 0 \\ 1 & \frac{d}{dx} \end{bmatrix} \begin{bmatrix} Q_y \\ M_z \end{bmatrix} + \begin{bmatrix} q_y \\ m_z \end{bmatrix} = \begin{bmatrix} 0 \\ 0 \end{bmatrix}$	$\mathcal{L}_1^T s^* + b^* = 0$
PDE	
$\frac{d}{dx} \left[k_s G A \left(\frac{du_y}{dx} - \phi_z \right) \right] + q_y = 0$ $\frac{d}{dx} \left(E I_z \frac{d\phi_z}{dx} \right) + k_s G A \left(\frac{du_y}{dx} - \phi_z \right) + m_z = 0,$	$\mathcal{L}_1^T D^* \mathcal{L}_1 * u + b^* = 0$

Solution of the coupled DEs for a Timoshenko beam with kAG, EI, q, m = const

```
(%i14)  eqn_1: kAG*'diff(u(x),x,2) = -q+kAG*'diff(phi(x),x)$
        eqn_2: kAG*'diff(u(x),x) = -m-EI*'diff(phi(x),x,2)+kAG*phi(x)$

        sol : desolve([eqn_1, eqn_2], [u(x), phi(x)])$

        print(" ")$
        print("Equations:")$
        print(" ", eqn_1)$
        print(" ")$
        print(" ", eqn_2)$
        print(" ")$
        print(" ")$
        print("Analytical Solution:")$
        print(ratsimp(sol[1]))$
        print(" ")$
        print(ratsimp(sol[2]))$
```

Equations:

$$kAG \left(\frac{d^2}{dx^2} u(x) \right) = kAG \left(\frac{d}{dx} \phi(x) \right) - q$$

$$kAG \left(\frac{d}{dx} u(x) \right) = -EI \left(\frac{d^2}{dx^2} \phi(x) \right) + kAG \, \phi(x) - m$$

Analytical Solution:

$$u(x) = - \frac{ \left(4(kAG)^2 x^3 - 24EI \, kAG x\right) \left(\frac{d}{dx} u(x) \big|_{x=0} \right) }{24EI \, kAG} $$

$$\begin{aligned}
u(x) = -\frac{1}{24EI\,kAG}\Big[&\left(4(kAG)^2 x^3 - 24EI\,kAGx\right)\left(\tfrac{d}{dx}u(x)\big|_{x=0}\right) \\
&-12EI\,kAGx^2\left(\tfrac{d}{dx}\mathrm{phi}(x)\big|_{x=0}\right) - kAGqx^4 \\
&+ \left(-4\,\mathrm{phi}(0)(kAG)^2 + 4kAGm\right)x^3 + 12EIqx^2 - 24\,u(0)EI\,kAG \Big]
\end{aligned}$$

$$\begin{aligned}
\mathrm{phi}(x) = -\frac{1}{6EI}\Big[&3kAGx^2\left(\tfrac{d}{dx}u(x)\big|_{x=0}\right) - 6EIx\left(\tfrac{d}{dx}\mathrm{phi}(x)\big|_{x=0}\right) - qx^3 \\
&+ \left(-3\,\mathrm{phi}(0)kAG + 3m\right)x^2 - 6\,\mathrm{phi}(0)EI \Big]
\end{aligned}$$

Fig. 3.8 Solution of the coupled system of differential equations for the Timoshenko beams based on the computer algebra system Maxima

Under the assumption of constant material (E, G) and geometric (I_z, A, k_s) properties, the system of differential equations in Table 3.5 can be solved by using a computer algebra system for constant distributed loads ($q_y(x) = q_0 = $ const. and $m_z(x) = m_0 = $ const.) to obtain the general analytical solution of the problem, see Fig. 3.8.

The general analytical solution provided in Fig. 3.8, i.e.,

$$u_y(x) = -\frac{\begin{aligned}&\left(4(k_sAG)^2x^3 - 24EI_z k_sAGx\right)\frac{du_y(0)}{dx}\\&-12EI_z k_sAGx^2\frac{d\phi_z(0)}{dx} - k_sAGq_0x^4\\&+\left(-4\phi_z(0)(kAG)^2 + 4k_sAGm_0\right)x^3 + 12EI_zq_0x^2\\&\qquad\qquad\qquad\qquad\qquad\quad - 24u_y(0)EI_z k_sAG\end{aligned}}{24EI_z k_sAG}, \quad (3.41)$$

$$\phi_z(x) = -\frac{3k_sAGx^2\frac{du_y(0)}{dx} - 6EI_zx\,\frac{d\phi_z(0)}{dx} - q_0x^3}{6EI_z} + (-3\phi_z(0)k_sAG + 3m_0)\,x^2 - 6\phi_z(0)EI_z}{6EI_z}, \quad (3.42)$$

where $\frac{du_y(0)}{dx}$, $\frac{d\phi_z(0)}{dx}$, $u_y(0)$, and $\phi_z(0)$ are the four boundary values to adjust the general solution to a particular problem, can be represented in a slightly different way based on a different set of constants:

$$
\begin{aligned}
u_y(x) = \frac{1}{EI_z}\Bigg(&\frac{q_0x^4}{24} + \frac{x^3}{6}\underbrace{\left[-k_sAG\frac{du_y(0)}{dx} + k_sAG\phi_z(0)\right]}_{c_1} + \\
&+ \frac{x^2}{2}\underbrace{\left[+EI_z\frac{d\phi_z(0)}{dx} - \frac{EI_zq_0}{k_sAG}\right]}_{c_2} + \underbrace{EI_z\frac{du_y(0)}{dx}}_{c_3}x + \\
&+ \underbrace{EI_zu_y(0)}_{c_4}\Bigg) - \frac{m_0x^3}{6EI_z},
\end{aligned}
\quad (3.43)
$$

$$
\begin{aligned}
\phi_z(x) = +\frac{1}{EI_z}\Bigg(&\frac{q_0x^3}{3} + \frac{x^2}{2}\underbrace{\left[-k_sAG\frac{du_y(0)}{dx} + k_sAG\phi_z(0)\right]}_{c_1} + \\
&+ x\underbrace{\left[+EI_z\frac{d\phi_z(0)}{dx} - \frac{EI_zq_0}{k_sAG}\right]}_{c_2} + \underbrace{EI_z\frac{du_y(0)}{dx}}_{c_3}\Bigg) \\
&+ \frac{q_0x}{k_sAG} + \underbrace{\left[-\frac{du_y(0)}{dx} + \phi_z(0)\right]}_{\frac{c_1}{k_sAG}} - \frac{m_0x^2}{2EI_z}.
\end{aligned}
\quad (3.44)
$$

The general expressions for the internal bending moment and shear force distributions can be obtained under consideration of Eqs. (3.23) and (3.24). Thus, we can finally state the general solution for constant material and geometrical properties as well as constant distributed loads as [8]:

$$u_y(x) = \frac{1}{EI_z}\left(\frac{q_0 x^4}{24} + c_1\frac{x^3}{6} + c_2\frac{x^2}{2} + c_3 x + c_4\right) - \frac{m_0 x^3}{6EI_z}, \tag{3.45}$$

$$\phi_z(x) = +\frac{1}{EI_z}\left(\frac{q_0 x^3}{6} + c_1\frac{x^2}{2} + c_2 x + c_3\right) + \frac{q_0 x}{k_s AG} + \frac{c_1}{k_s AG} - \frac{m_0 x^2}{2EI_z}, \tag{3.46}$$

$$M_z(x) = +\left(\frac{q_0 x^2}{2} + c_1 x + c_2\right) + \frac{q_0 EI_z}{k_s AG} - m_0 x, \tag{3.47}$$

$$Q_y(x) = -\left(q_0 x + c_1\right), \tag{3.48}$$

where the four constants of integration $c_i\,(i = 1, \ldots, 4)$ must be determined based on the boundary conditions.

3.3 Deformation in the x-z Plane

For the derivation of the equation for the shear strain in the x-z plane, the infinitesimal rectangular beam element $ABCD$, shown in Fig. 3.9, is considered, which deforms under the influence of a pure shear stress. Here, a change of the angle of the original right angles as well as a change in the lengths of the edges occurs.

The deformation of the point A can be described based on the displacement fields $u_x(x, z)$ and $u_z(x, z)$. These two functions of *two* variables can be expanded into Taylor's series[4] of first order around point A to approximately calculate the deformations of the points B and D:

$$u_{x,B} = u_x(x + dx, z) = u_x(x, z) + \frac{\partial u_x(x, z)}{\partial x}dx + \frac{\partial u_x(x, z)}{\partial z}dz, \tag{3.49}$$

$$u_{z,B} = u_z(x + dx, z) = u_z(x, z) + \frac{\partial u_z(x, z)}{\partial x}dx + \frac{\partial u_z(x, z)}{\partial z}dz, \tag{3.50}$$

---

[4]For a function $f(x, z)$ of two variables usually a Taylor's series expansion of first-order is formulated around the point (x_0, z_0) as follows: $f(x, z) = f(x_0 + dx, z_0 + dz) \approx f(x_0, z_0) + \left(\frac{\partial f}{\partial x}\right)_{x_0, z_0} \times (x - x_0) + \left(\frac{\partial f}{\partial z}\right)_{x_0, z_0} \times (z - z_0)$.

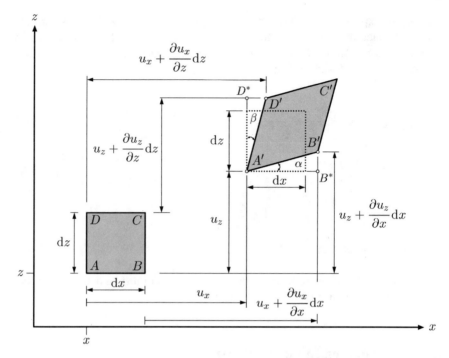

Fig. 3.9 Definition of the shear strain γ_{xz} in the x-z plane at an infinitesimal beam element

or alternatively

$$u_{x,D} = u_x(x, z + dz) = u_x(x, z) + \frac{\partial u_x(x, z)}{\partial x}dx + \frac{\partial u_x(x, z)}{\partial z}dz , \qquad (3.51)$$

$$u_{z,D} = u_z(x, z + dz) = u_z(x, z) + \frac{\partial u_z(x, z)}{\partial x}dx + \frac{\partial u_z(x, z)}{\partial z}dz . \qquad (3.52)$$

In Eqs. (3.49) up to (3.52), $u_x(x, z)$ and $u_z(x, z)$ represent the so-called rigid-body displacements, which do not cause a deformation. If one considers that point B has the coordinates $(x + dx, z)$ and D the coordinates $(x, z + dz)$, the following results:

$$u_{x,B} = u_x(x, z) + \frac{\partial u_x(x, z)}{\partial x}dx , \qquad (3.53)$$

$$u_{z,B} = u_z(x, z) + \frac{\partial u_z(x, z)}{\partial x}dx , \qquad (3.54)$$

or alternatively

$$u_{x,D} = u_x(x, z) + \frac{\partial u_x(x, z)}{\partial z} dz, \tag{3.55}$$

$$u_{z,D} = u_z(x, z) + \frac{\partial u_z(x, z)}{\partial z} dz. \tag{3.56}$$

The total shear strain γ_{xz} of the deformed beam element $A'B'C'D'$ results, according to Fig. 3.9, from the sum of the angles α and β. The two angles can be identified in the rectangle, which is deformed to a rhombus. Under consideration of the two right-angled triangles $A'D^*D'$ and $A'B^*B'$, these two angles can be expressed as:

$$\tan \alpha = \frac{\dfrac{\partial u_z(x, z)}{\partial x} dx}{dx + \dfrac{\partial u_x(x, z)}{\partial x} dx} \quad \text{and} \quad \tan \beta = \frac{\dfrac{\partial u_x(x, z)}{\partial z} dz}{dz + \dfrac{\partial u_z(x, z)}{\partial z} dz}. \tag{3.57}$$

It holds approximately for small deformations that $\tan \alpha \approx \alpha$ and $\tan \beta \approx \beta$ or alternatively $\frac{\partial u_x}{\partial x} \ll 1$ and $\frac{\partial u_z}{\partial z} \ll 1$, so that the following expression results for the shear strain in the x-z plane:

$$\gamma_{xz} = \alpha + \beta \approx \frac{\partial u_z(x, z)}{\partial x} + \frac{\partial u_x(x, z)}{\partial z}. \tag{3.58}$$

This total change of the angle is also called the engineering shear strain. In contrast to this, the expression $\varepsilon_{xz} = \frac{1}{2} \gamma_{xz} = \frac{1}{2}(\frac{\partial u_z}{\partial x} + \frac{\partial u_x}{\partial z})$ is known as the tensorial definition (tensor shear strain) in the literature [18]. Due to the symmetry of the strain tensor, the identity $\gamma_{ij} = \gamma_{ji}$ applies to the tensor elements outside the main diagonal.

The algebraic sign of the shear strain needs to be explained in the following with the help of Fig. 3.10 for the special case that only one shear force acts in parallel to the z-axis. If a shear force acts in the direction of the positive z-axis at the right-hand face—hence a positive shear force distribution is being assumed at this point —, according to Fig. 3.10a under consideration of Eq. (3.58) a positive shear strain results. In a similar way, a negative shear force distribution leads to a negative shear strain according to Fig. 3.10b.

It has already been mentioned in Sect. 2.1 that the shear stress distribution is variable over the cross-section. As an example, the parabolic shear stress distribution was illustrated over a rectangular cross section in Fig. 2.2. Based on Hooke's law for a one-dimensional shear stress state, it can be derived that the shear strain has to exhibit a corresponding parabolic course. From the shear stress distribution in the

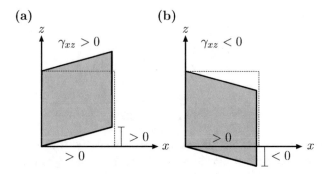

Fig. 3.10 Definition of a **a** positive and **b** negative shear strain in the x-z plane

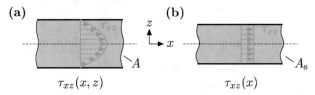

Fig. 3.11 Shear stress distribution in the x-z plane: **a** real distribution for a rectangular cross section and **b** Timoshenko's approximation

cross-sectional area at location x of the beam,[5] one receives the acting shear force through integration over the cross section as:

$$Q_z(x) = \int_A \tau_{xz}(x, y, z) \, dA . \tag{3.59}$$

However, to simplify the problem, it is assumed for the Timoshenko beam that an equivalent *constant* shear stress and strain act at location x, see Fig. 3.11:

$$\tau_{xz}(x, y, z) \rightarrow \tau_{xz}(x) . \tag{3.60}$$

This constant shear stress results from the shear force, which acts in an equivalent cross-sectional area, the so-called shear area A_s:

$$\tau_{xz}(x) = \frac{Q_z(x)}{A_s} , \tag{3.61}$$

---

[5]A closer analysis of the shear stress distribution in the cross-sectional area shows that the shear stress does not just alter over the height of the beam but also through the width of the beam. If the width of the beam is small when compared to the height, only a small change along the width occurs and one can assume in the first approximation a constant shear stress throughout the width: $\tau_{xz}(y, z) \rightarrow \tau_{xz}(z)$. See for example [3, 17].

Table 3.6 Comparison of shear correction factor values for a rectangular cross section based on different approaches

k_s	Comment	Reference
$\frac{2}{3}$	–	[14, 16]
$0.833\left(=\frac{5}{6}\right)$	$\nu = 0.0$	[4]
0.850	$\nu = 0.3$	
0.870	$\nu = 0.5$	

whereupon the relation between the shear area A_s and the actual cross-sectional area A_s is referred to as the shear correction factor k_s:

$$k_s = \frac{A_s}{A}. \tag{3.62}$$

Different assumptions can be made to calculate the shear correction factor [4]. As an example, it can be demanded [2] that the elastic strain energy of the equivalent shear stress has to be identical with the energy, which results from the acting shear stress distribution in the actual cross-sectional-area. A comparison for a rectangular cross section is presented in Table 3.6.

Different geometric characteristics of simple geometric cross-sections—including the shear correction factor[6]—are collected in Table 3.7 [5, 19]. Further details regarding the shear correction factor for arbitrary cross-sections can be taken from [6].

It is obvious that the equivalent constant shear stress can alter along the center line of the beam, in case the shear force along the center line of the beam changes. The attribute 'constant' thus just refers to the cross-sectional area at location x and the equivalent constant shear stress is therefore in general a function of the coordinate of length for the Timoshenko beam:

$$\tau_{xz} = \tau_{xz}(x). \tag{3.63}$$

The so-called Timoshenko beam can be generated by superposing a shear deformation on a Euler–Bernoulli beam according to Fig. 3.12.

One can see that Bernoulli's hypothesis is partly no longer fulfilled for the Timoshenko beam: Plane cross sections remain plane after the deformation. However, a cross section which stood at right angles on the beam axis before the deformation is not at right angles on the beam axis after the deformation. If the demand for planeness of the cross sections is also given up, one reaches theories of higher-order [7, 11, 12], at which, for example, a parabolic course of the shear strain and stress in the displacement field are considered, see Fig. 3.13. Therefore, a shear correction factor is not required for these theories of higher-order.

---

[6]It should be noted that the so-called form factor for shear is also known in the literature. This results as the reciprocal of the shear correction factor.

Table 3.7 Characteristics of different cross sections in the y-z plane. I_y: axial second moment of area; A: cross-sectional area; k_s: shear correction factor. Adapted from [19]

Cross-section	I_y	A	k_s
	$\dfrac{\pi R^4}{4}$	πR^2	$\dfrac{9}{10}$
	$\pi R^3 t$	$2\pi R t$	0.5
	$\dfrac{bh^3}{12}$	hb	$\dfrac{5}{6}$
	$\dfrac{h^2}{6}(ht_w + 3bt_f)$	$2(bt_f + ht_w)$	$\dfrac{2ht_w}{A}$
	$\dfrac{h^2}{12}(ht_w + 6bt_f)$	$ht_w + 2bt_f$	$\dfrac{ht_w}{A}$

3.3.1 Kinematics

According to the alternative derivation in Sect. 2.3.1, the kinematics relation can also be derived for the beam with shear action, by considering the angle ϕ_y instead of the angle φ_y, see Figs. 3.12c and 3.14.

Following an equivalent procedure as in Sect. 2.3.1, the corresponding relationships are obtained:

$$\sin\phi_y = \frac{u_x}{z} \approx \phi_y \quad \text{or} \quad u_x = +z\phi_y, \tag{3.64}$$

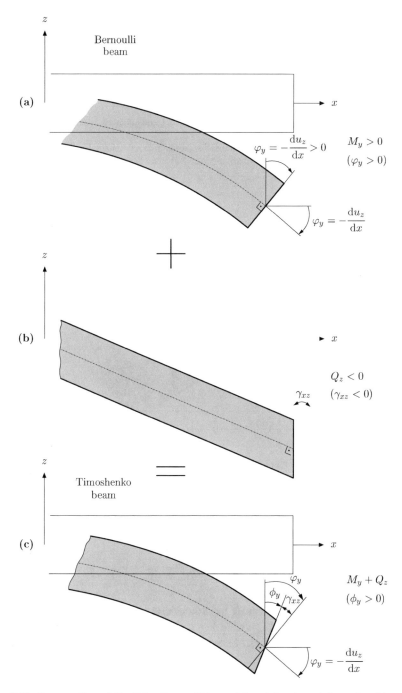

Fig. 3.12 Superposition of the Euler–Bernoulli beam (**a**) and the shear deformation (**b**) to the Timoshenko beam (**c**) in the x-z plane. See Fig. 3.10 for clarification on the sign of γ_{xz}. Note that the deformation is exaggerated for better illustration

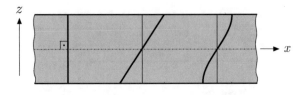

Bernoulli   Timoshenko   Levinson

Fig. 3.13 Deformation of originally plane cross sections in the x-z plane for the Euler–Bernoulli beam (left), the Timoshenko beam (middle), and the Levinson beam (right) [13]

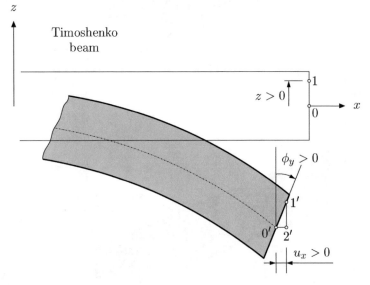

Fig. 3.14 Derivation of the kinematics relation in the x-z plane. Note that the deformation is exaggerated for better illustration

wherefrom, via the general relation for the strain, meaning $\varepsilon_x = \mathrm{d}u_x/\mathrm{d}x$, the kinematics relation results through differentiation with respect to the x-coordinate:

$$\varepsilon_x = +z\frac{\mathrm{d}\phi_y}{\mathrm{d}x}. \tag{3.65}$$

Note that $\phi_y \to \varphi_y = -\frac{\mathrm{d}u_z}{\mathrm{d}x}$ results from neglecting the shear deformation and a relation according to Eq. (2.56) results as a special case. Furthermore, the following relation between the angles can be derived from Fig. 3.12c

$$\phi_y = \varphi_y + \gamma_{xz} = -\frac{\mathrm{d}u_z}{\mathrm{d}x} + \gamma_{xz}, \tag{3.66}$$

which complements the set of the kinematics relations. It needs to be remarked that at this point the so-called bending line was considered. Therefore, the displacement field u_z is only a function of *one* variable: $u_z = u_z(x)$.

3.3.2 Constitutive Equation

For the consideration of the constitutive relation, Hooke's law for a one-dimensional normal stress state and for a one-dimensional shear stress state is used:

$$\sigma_x = E\varepsilon_x, \tag{3.67}$$

$$\tau_{xz} = G\gamma_{xz}, \tag{3.68}$$

whereupon the shear modulus G can be calculated for isotropic materials based on the Young's modulus E and the Poisson's ratio ν as:

$$G = \frac{E}{2(1+\nu)}. \tag{3.69}$$

According to the equilibrium configuration of Fig. 2.19 and Eq. (2.62), the relation between the internal moment and the bending stress can be used for the Timoshenko beam as follows:

$$dM_y = (+z)(+\sigma_x)dA, \tag{3.70}$$

or alternatively after integration under the consideration of the constitutive equation (3.67) and the kinematics relation (3.65):

$$M_y(x) = +EI_y\frac{d\phi_y(x)}{dx}. \tag{3.71}$$

The relation between shear force and cross-sectional rotation results from the equilibrium equation (3.76) as:

$$Q_z(x) = +\frac{dM_y(x)}{dx} + m_y(x) = +EI_y\frac{d^2\phi_y(x)}{dx^2} + m_y(x). \tag{3.72}$$

Alternatively, the internal shear force results from Eq. (3.61) and the application of the constitutive relation (3.68) as:

$$Q_z(x) = \tau_{xz}A_s = k_s AG\gamma_{xz}(x), \tag{3.73}$$

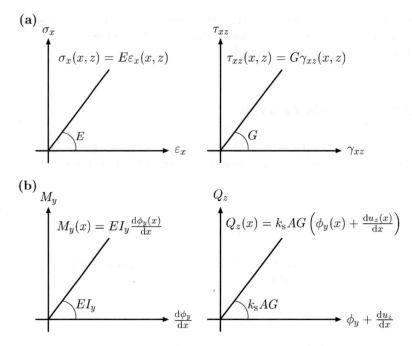

Fig. 3.15 Formulation of the constitutive law based on **a** classical stress-strain and **b** generalized-stress-generalized-strain relations (x-z plane)

or under consideration of the kinematics relation (3.66) as a solely function of the deformations $u_z(x)$ and $\phi_y(x)$:

$$Q_z(x) = k_s AG \left(\frac{\mathrm{d}u_z(x)}{\mathrm{d}x} + \phi_y(x) \right) . \tag{3.74}$$

Similar to Sect. 2.3.2, one can use the stress resultants to formulate the constitutive law based on the generalized stresses $s = \begin{bmatrix} M_y, & Q_z \end{bmatrix}^{\mathrm{T}}$ and generalized strains $e = \begin{bmatrix} \frac{\mathrm{d}\phi_y}{\mathrm{d}x}, & \phi_y + \frac{\mathrm{d}u_z}{\mathrm{d}x} \end{bmatrix}^{\mathrm{T}}$, see Fig. 3.15. The generalized quantities reveal the advantage that they are not a function of the vertical coordinate.

3.3.3 Equilibrium

The derivation of the equilibrium condition for the Timoshenko beam is identical with the derivation for the Euler–Bernoulli beam according to Sect. 2.3.3:

Table 3.8 Elementary basic equations for the bending of a Timoshenko beam in the *x-z* plane (*x*-axis: right facing; *z*-axis: upward facing)

Relation	Equation
Kinematics	$\varepsilon_x(x, z) = +z\dfrac{\mathrm{d}\phi_y(x)}{\mathrm{d}x}$ and $\phi_y(x) = -\dfrac{\mathrm{d}u_z(x)}{\mathrm{d}x} + \gamma_{xz}(x)$
Equilibrium	$\dfrac{\mathrm{d}Q_z(x)}{\mathrm{d}x} = -q_z(x)$ $\dfrac{\mathrm{d}M_y(x)}{\mathrm{d}x} = +Q_z(x) - m_y(x)$
Constitution	$\sigma_x(x, z) = E\varepsilon_x(x, z)$ and $\tau_{xz}(x) = G\gamma_{xz}(x)$

$$\frac{\mathrm{d}Q_z(x)}{\mathrm{d}x} = -q_z(x)\,, \tag{3.75}$$

$$\frac{\mathrm{d}M_y(x)}{\mathrm{d}x} = +Q_z(x) - m_y(x)\,. \tag{3.76}$$

Before looking in more detail at the differential equations of the bending line, let us summarize the basic equations for the Timoshenko beam in Table 3.8. Note that the normal stress and normal strain are functions of both spatial coordinates, i.e. *x* and *z*. However, the shear stress and shear strain are only dependent on the *x*-coordinate, since an equivalent *constant* shear stress has been introduced over the cross section as an approximation of the Timoshenko beam theory.

3.3.4 Differential Equation

Within the previous section, the relation between the internal moment and the cross-sectional rotation was derived from the normal stress distribution with the help of Hooke's law, see Eq. (3.71). Differentiation of this relation with respect to the *x*-coordinate leads to the following expression

$$\frac{\mathrm{d}M_y(x)}{\mathrm{d}x} = \frac{\mathrm{d}}{\mathrm{d}x}\left(EI_y\frac{\mathrm{d}\phi_y(x)}{\mathrm{d}x}\right)\,, \tag{3.77}$$

which can be transformed with the help of the equilibrium relation (3.76), the constitutive equation (3.68), and the relation for the shear stress according to (3.61) and (3.62) to

$$\frac{\mathrm{d}}{\mathrm{d}x}\left(EI_y\frac{\mathrm{d}\phi_y(x)}{\mathrm{d}x}\right) = +k_s GA\gamma_{xz}(x) - m_y(x)\,. \tag{3.78}$$

If the kinematics relation (3.66) is considered in the last equation, the so-called bending differential equation results in:

$$\frac{d}{dx}\left(EI_y\frac{d\phi_y(x)}{dx}\right) - k_sGA\left(\frac{du_z(x)}{dx} + \phi_y(x)\right) = -m_y(x). \tag{3.79}$$

Considering the shear stress according to Eqs. (3.61) and (3.62) in the expression of Hooke's law according to (3.68), one obtains

$$Q_z(x) = k_sAG\gamma_{xz}(x). \tag{3.80}$$

Introducing the equilibrium relation (3.76) and the kinematics relation (3.66) in the last equation gives:

$$\frac{dM_y(x)}{dx} + m_y(x) = +k_sAG\left(\frac{du_z(x)}{dx} + \phi_y(x)\right). \tag{3.81}$$

After differentiation and the consideration of the equilibrium relations according to Eqs. (3.75) and (3.76), the so-called shear differential equation results finally in:

$$\frac{d}{dx}\left[k_sAG\left(\frac{du_z(x)}{dx} + \phi_y(x)\right)\right] = -q_z(x). \tag{3.82}$$

Therefore, the shear flexible Timoshenko beam is described through the following two coupled differential equations of second order:

$$\frac{d}{dx}\left(EI_y\frac{d\phi_y(x)}{dx}\right) - k_sAG\left(\frac{du_z(x)}{dx} + \phi_y(x)\right) = -m_y(x), \tag{3.83}$$

$$\frac{d}{dx}\left[k_sAG\left(\frac{du_z(x)}{dx} + \phi_y(x)\right)\right] = -q_z(x). \tag{3.84}$$

This system contains two unknown functions, namely the deflection $u_z(x)$ and the cross-sectional rotation $\phi_y(x)$. Boundary conditions must be formulated for both functions to be able to solve the system of differential equations for a specific problem.

Different formulations of these coupled differential equations are collected in Table 3.9 where different types of loadings, geometry and bedding are differentiated. The last case in Table 3.9 refers again to the elastic or Winkler foundation of a beam, [20]. The elastic foundation modulus k has in the case of beams the unit of force per unit area.

A single-equation description for the Timoshenko beam can be obtained under the assumption of constant material (E, G) and geometrical (I_z, A, k_s) properties: Rearranging and two-times differentiation of Eq. (3.84) gives:

Table 3.9 Different formulations of the partial differential equation for a Timoshenko beam in the *x*-*z* plane (*x*-axis: right facing; *z*-axis: upward facing)

Configuration	Partial differential equation
$E, I_y, A, G, k_{\mathrm{s}}$	$EI_y \dfrac{d^2\phi_y}{dx^2} - k_{\mathrm{s}}GA\left(\dfrac{du_z}{dx} + \phi_y\right) = 0$ $k_{\mathrm{s}}GA\left(\dfrac{d^2 u_z}{dx^2} + \dfrac{d\phi_y}{dx}\right) = 0$
$E(x), I_y(x)$ $k_{\mathrm{s}}(x), A(x), G(x)$	$\dfrac{d}{dx}\left(E(x)I_y(x)\dfrac{d\phi_y}{dx}\right) -$ $k_{\mathrm{s}}(x)G(x)A(x)\left(\dfrac{du_z}{dx} + \phi_y\right) = 0$ $\dfrac{d}{dx}\left[k_{\mathrm{s}}(x)G(x)A(x)\left(\dfrac{du_z}{dx} + \phi_y\right)\right] = 0$
$q_z(x)$	$EI_y \dfrac{d^2\phi_y}{dx^2} - k_{\mathrm{s}}GA\left(\dfrac{du_z}{dx} + \phi_y\right) = 0$ $k_{\mathrm{s}}GA\left(\dfrac{d^2 u_z}{dx^2} + \dfrac{d\phi_y}{dx}\right) = -q_z(x)$
$m_y(x)$	$EI_y \dfrac{d^2\phi_y}{dx^2} - k_{\mathrm{s}}GA\left(\dfrac{du_z}{dx} + \phi_y\right) = -m_y(x)$ $k_{\mathrm{s}}GA\left(\dfrac{d^2 u_z}{dx^2} + \dfrac{d\phi_y}{dx}\right) = 0$
$k(x)$	$EI_y \dfrac{d^2\phi_y}{dx^2} - k_{\mathrm{s}}GA\left(\dfrac{du_z}{dx} + \phi_y\right) = 0$ $k_{\mathrm{s}}GA\left(\dfrac{d^2 u_z}{dx^2} + \dfrac{d\phi_y}{dx}\right) = k(x)u_z$

$$\frac{d\phi_y(x)}{dx} = -\frac{d^2 u_z(x)}{dx^2} - \frac{q_z(x)}{k_{\mathrm{s}}GA}, \tag{3.85}$$

$$\frac{d^3\phi_y(x)}{dx^3} = -\frac{d^4 u_z(x)}{dx^4} - \frac{d^2 q_z(x)}{k_{\mathrm{s}}GA\, dx^2}. \tag{3.86}$$

One-time differentiation of Eq. (3.83) gives:

$$EI_y \frac{d^3\phi_y(x)}{dx^3} - k_{\mathrm{s}}AG\left(\frac{d^2 u_z(x)}{dx^2} + \frac{d\phi_y(x)}{dx}\right) = -\frac{dm_y(x)}{dx}. \tag{3.87}$$

Table 3.10 Different formulations of the basic equations for a Timoshenko beam (bending in the x-z plane; x-axis along the principal beam axis). E: Young's modulus; G: shear modulus; A: cross-sectional area; I_y: second moment of area; k_s: shear correction factor; q_z: length-specific distributed force; m_y: length-specific distributed moment; e: generalized strains; s^*: generalized stresses

Specific formulation	General formulation [1]
Kinematics	
$\begin{bmatrix} \frac{du_z}{dx} + \phi_y \\ \frac{d\phi_y}{dx} \end{bmatrix} = \begin{bmatrix} \frac{d}{dx} & 1 \\ 0 & \frac{d}{dx} \end{bmatrix} \begin{bmatrix} u_z \\ \phi_y \end{bmatrix}$	$e = \mathcal{L}_1 u$
Constitution	
$\begin{bmatrix} Q_z \\ M_y \end{bmatrix} = \begin{bmatrix} k_s A G & 0 \\ 0 & E I_y \end{bmatrix} \begin{bmatrix} \frac{du_z}{dx} + \phi_y \\ \frac{d\phi_y}{dx} \end{bmatrix}$	$s^* = D^* e$
Equilibrium	
$\begin{bmatrix} \frac{d}{dx} & 0 \\ -1 & \frac{d}{dx} \end{bmatrix} \begin{bmatrix} Q_z \\ M_y \end{bmatrix} + \begin{bmatrix} q_z \\ +m_y \end{bmatrix} = \begin{bmatrix} 0 \\ 0 \end{bmatrix}$	$\mathcal{L}_{1*}^T s^* + b^* = 0$
PDE	
$\frac{d}{dx}\left[k_s G A \left(\frac{du_z}{dx} + \phi_y \right) \right] + q_z = 0$ $\frac{d}{dx}\left(E I_y \frac{d\phi_y}{dx} \right) - k_s G A \left(\frac{du_z}{dx} + \phi_y \right) + m_y = 0,$	$\mathcal{L}_{1*}^T D^* \mathcal{L}_1 u + b^* = 0$

Inserting Eq. (3.85) into (3.87) and consideration of (3.86) gives finally the following expression:

$$E I_y \frac{d^4 u_z(x)}{dx^4} = q_z(x) + \frac{dm_y(x)}{dx} - \frac{E I_y}{k_s A G} \frac{d^2 q_z(x)}{dx^2}. \qquad (3.88)$$

The last equation reduces for shear-rigid beams, i.e. $k_s A G \to \infty$, to the classical Euler–Bernoulli formulation as given in Table 2.8.

If we replace in the previous formulations the first-order derivative, i.e. $\frac{d\cdots}{dx}$, by a formal operator symbol, i.e. the $\mathcal{L}_1$–matrix, then the basic equations of the Timoshenko beam can be stated in a more formal way as given in Table 3.10. Such a matrix formulation is suitable for the derivation of the principal finite element equation based on the weighted residual method [9, 10].

Under the assumption of constant material (E, G) and geometric (I_y, A, k_s) properties, the system of differential equations in Table 3.10 can be solved by using a computer algebra system for constant distributed loads ($q_z(x) = q_0 = $ const. and $m_y(x) = m_0 = $ const.) to obtain the general analytical solution of the problem, see Fig. 3.16.

Solution of the coupled DEs for a Timoshenko beam with kAG, EI, q, m = const

(%i14) eqn_1: -kAG*'diff(u(x),x,2) = q+kAG*'diff(phi(x),x)$
        eqn_2: -kAG*'diff(u(x),x) = -m-EI*'diff(phi(x),x,2)+kAG*phi(x)$

        sol : desolve([eqn_1, eqn_2], [u(x), phi(x)])$

        print(" ")$
        print("Equations:")$
        print(" ", eqn_1)$
        print(" ")$
        print(" ", eqn_2)$
        print(" ")$
        print(" ")$
        print("Analytical Solution:")$
        print(ratsimp(sol[1]))$
        print(" ")$
        print(ratsimp(sol[2]))$

Equations:

$$-kAG\left(\frac{\mathrm{d}^2}{\mathrm{d}x^2}\,u(x)\right) = kAG\left(\frac{\mathrm{d}}{\mathrm{d}x}\phi(x)\right) + q$$

$$-kAG\left(\frac{\mathrm{d}}{\mathrm{d}x}\,u(x)\right) = -EI\left(\frac{\mathrm{d}^2}{\mathrm{d}x^2}\phi(x)\right) + kAG\,\phi(x) - m$$

Analytical Solution:

$$u(x) = -\frac{\begin{array}{c}\left(4(kAG)^2x^3 - 24EI\,kAGx\right)\left(\frac{\mathrm{d}}{\mathrm{d}x}\,u(x)\big|_{x=0}\right) \\ +12EI\,kAGx^2\left(\frac{\mathrm{d}}{\mathrm{d}x}\,\mathrm{phi}(x)\big|_{x=0}\right) - kAGqx^4 \\ + \left(4\,\mathrm{phi}(0)(kAG)^2 - 4kAGm\right)x^3 + 12EIqx^2 - 24\,u(0)EI\,kAG\end{array}}{24EI\,kAG}$$

$$\mathrm{phi}(x) = \frac{\begin{array}{c}3kAGx^2\left(\frac{\mathrm{d}}{\mathrm{d}x}\,u(x)\big|_{x=0}\right) + 6EIx\left(\frac{\mathrm{d}}{\mathrm{d}x}\,\mathrm{phi}(x)\big|_{x=0}\right) - qx^3 \\ + \left(3\,\mathrm{phi}(0)kAG - 3m\right)x^2 + 6\,\mathrm{phi}(0)EI\end{array}}{6EI}$$

Fig. 3.16 Solution of the coupled system of differential equations for the Timoshenko beams based on the computer algebra system Maxima

The general analytical solution provided in Fig. 3.16, i.e.,

$$u_z(x) = -\frac{\begin{array}{c}\left(4(k_sAG)^2x^3 - 24EI_y\,k_sAGx\right)\frac{\mathrm{d}u_z(0)}{\mathrm{d}x} \\ +12EI_y\,k_sAGx^2\frac{\mathrm{d}\phi_y(0)}{\mathrm{d}x} - k_sAGq_0x^4 \\ + \left(4\phi_y(0)(kAG)^2 - 4k_sAGm_0\right)x^3 + 12EI_yq_0x^2 \\ - 24u_z(0)EI_y\,k_sAG\end{array}}{24EI_y\,k_sAG}, \quad (3.89)$$

$$\phi_y(x) = \frac{3k_sAGx^2\frac{du_z(0)}{dx} + 6EI_yx\frac{d\phi_y(0)}{dx} - q_0x^3}{+\left(3\phi_y(0)k_sAG - 3m_0\right)x^2 + 6\phi_y(0)EI_y}{6EI_y},$$ (3.90)

where $\frac{du_z(0)}{dx}$, $\frac{d\phi_y(0)}{dx}$, $u_z(0)$, and $\phi_y(0)$ are the four boundary values to adjust the general solution to a particular problem, can be represented in a slightly different way based on a different set of constants:

$$u_z(x) = \frac{1}{EI_y}\Bigg(\frac{q_0x^4}{24} + \frac{x^3}{6}\underbrace{\left[-k_sAG\frac{du_z(0)}{dx} - k_sAG\phi_y(0)\right]}_{c_1} +$$

$$+ \frac{x^2}{2}\underbrace{\left[-EI_y\frac{d\phi_y(0)}{dx} - \frac{EI_yq_0}{k_sAG}\right]}_{c_2} + \underbrace{EI_y\frac{du_z(0)}{dx}}_{c_3}x +$$

$$+ \underbrace{EI_yu_z(0)}_{c_4}\Bigg) + \frac{m_0x^3}{6EI_y},$$ (3.91)

$$\phi_y(x) = -\frac{1}{EI_y}\Bigg(\frac{q_0x^3}{3} + \frac{x^2}{2}\underbrace{\left[-k_sAG\frac{du_z(0)}{dx} - k_sAG\phi_y(0)\right]}_{c_1} +$$

$$+ x\underbrace{\left[-EI_y\frac{d\phi_y(0)}{dx} - \frac{EI_yq_0}{k_sAG}\right]}_{c_2} + \underbrace{EI_y\frac{du_z(0)}{dx}}_{c_3}\Bigg)$$

$$- \frac{q_0x}{k_sAG} - \underbrace{\left[-\frac{du_z(0)}{dx} - \phi_y(0)\right]}_{\frac{c_1}{k_sAG}} - \frac{m_0x^2}{2EI_y}.$$ (3.92)

The general expressions for the internal bending moment and shear force distributions can be obtained under consideration of Eqs. (3.71) and (3.72). Thus, we can finally state the general solution for constant material and geometrical properties as well as constant distributed loads as:

$$u_z(x) = \frac{1}{EI_y}\left(\frac{q_0 x^4}{24} + c_1\frac{x^3}{6} + c_2\frac{x^2}{2} + c_3 x + c_4\right) + \frac{m_0 x^3}{6EI_y}, \qquad (3.93)$$

$$\phi_y(x) = -\frac{1}{EI_y}\left(\frac{q_0 x^3}{6} + c_1\frac{x^2}{2} + c_2 x + c_3\right) - \frac{q_0 x}{k_s AG} - \frac{c_1}{k_s AG} - \frac{m_0 x^2}{2EI_y}, \qquad (3.94)$$

$$M_y(x) = -\left(\frac{q_0 x^2}{2} + c_1 x + c_2\right) - \frac{q_0 EI_y}{k_s AG} - m_0 x, \qquad (3.95)$$

$$Q_z(x) = -(q_0 x + c_1), \qquad (3.96)$$

where the four constants of integration $c_i (i = 1, \ldots, 4)$ must be determined based on the boundary conditions.

3.4 Comparison of Both Planes

The comparison of the elementary basic equations for bending in the x-y and x-y plane are shown in Table 3.11. All the basic equations have the same structure but the different definition of a positive rotation yields to some different signs.

3.5 Unsymmetrical Bending in Both Planes

The line of reasoning presented in Sect. 2.4 is followed in the following. The total normal strain in the x-direction is obtained by simple superposition of its components from the bending contributions in the x-y and x-z plane (see Eqs. (3.17) and (3.65) for details), i.e.

$$\varepsilon_x(x, y, z) = \varepsilon_x^{x-y}(x, y) + \varepsilon_x^{x-z}(x, z). \qquad (3.97)$$

Thus, Hooke's law according to Eqs. (3.19) or (3.67) can be generalized to the following formulation:

$$\sigma_x(x, y, z) = E\varepsilon_x(x, y, z) = E\left(-y\frac{d\phi_z(x)}{dx} + z\frac{d\phi_y(x)}{dx}\right). \qquad (3.98)$$

Based on this stress distribution, the evaluation of the internal bending moments $M_z(x)$ and $M_y(x)$ (see Eqs. (3.22) and (3.70) for details) can be performed to obtain the following relations:

Table 3.11 Elementary basic equations for the bending of a Timoshenko beam in the x-y and x-z plane. The differential equations are given under the assumption of constant material (E, G) and geometrical (I, A, k_s) properties

Equation	x-y plane	x-z plane
Kinematics	$\varepsilon_x(x,y) = -y\dfrac{d\phi_z(x)}{dx}$	$\varepsilon_x(x,z) = +z\dfrac{d\phi_y(x)}{dx}$
	$\phi_z(x) = \dfrac{du_y(x)}{dx} - \gamma_{xy}(x)$	$\phi_y(x) = -\dfrac{du_z(x)}{dx} + \gamma_{xz}(x)$
Equilibrium	$\dfrac{dQ_y(x)}{dx} = -q_y(x)$	$\dfrac{dQ_z(x)}{dx} = -q_z(x)$
	$\dfrac{dM_z(x)}{dx} = -Q_y(x) - m_z(x)$	$\dfrac{dM_y(x)}{dx} = +Q_z(x) - m_y(x)$
Constitution	$\sigma_x(x,y) = E\varepsilon_x(x,y)$	$\sigma_x(x,z) = E\varepsilon_x(x,z)$
	$\tau_{xy}(x) = G\gamma_{xy}(x)$	$\tau_{xz}(x) = G\gamma_{xz}(x)$
	$M_z(x) = +EI_z\dfrac{d\phi_z(x)}{dx}$	$M_y(x) = +EI_y\dfrac{d\phi_y(x)}{dx}$
	$Q_y(x) = $	$Q_z(x) = $
	$k_sAG\left(\dfrac{du_y(x)}{dx} - \phi_z(x)\right)$	$k_sAG\left(\dfrac{du_z(x)}{dx} + \phi_y(x)\right)$
PDEs	$EI_z\dfrac{d^2\phi_z}{dx^2} +$	$EI_y\dfrac{d^2\phi_y}{dx^2} -$
	$k_sGA\left(\dfrac{du_y}{dx} - \phi_z\right)$	$k_sGA\left(\dfrac{du_z}{dx} + \phi_y\right)$
	$= -m_z(x)$	$= -m_y(x)$
	$k_sGA\left(\dfrac{d^2u_y}{dx^2} - \dfrac{d\phi_z}{dx}\right) =$	$k_sGA\left(\dfrac{d^2u_z}{dx^2} + \dfrac{d\phi_y}{dx}\right) =$
	$-q_y(x)$	$-q_z(x)$

$$M_z(x) = \int_A -y\sigma_x dA = +EI_z\frac{d\phi_z(x)}{dx} + EI_{yz}\frac{d\phi_y(x)}{dx}, \qquad (3.99)$$

$$M_y(x) = \int_A z\sigma_x dA = +EI_{yz}\frac{d\phi_z(x)}{dx} + EI_y\frac{d\phi_y(x)}{dx}, \qquad (3.100)$$

or for the internal shear forces $Q_y(x)$ and $Q_z(x)$ (see Eqs. (3.26) and (3.74) for details):

$$Q_y(x) = k_s A G \left(\frac{du_y(x)}{dx} - \phi_z(x) \right), \tag{3.101}$$

$$Q_z(x) = k_s A G \left(\frac{du_z(x)}{dx} + \phi_y(x) \right). \tag{3.102}$$

Equations (3.99)–(3.102) can be summarized in matrix form:

$$\begin{bmatrix} M_y(x) \\ M_z(x) \\ Q_y(x) \\ Q_z(x) \end{bmatrix} = \begin{bmatrix} EI_y & EI_{yz} & 0 & 0 \\ EI_{yz} & EI_z & 0 & 0 \\ 0 & 0 & k_s A G & 0 \\ 0 & 0 & 0 & k_s A G \end{bmatrix} \begin{bmatrix} \dfrac{d\phi_y(x)}{dx} \\ \dfrac{d\phi_z(x)}{dx} \\ \dfrac{du_y(x)}{dx} - \phi_z(x) \\ \dfrac{du_z(x)}{dx} + \phi_y(x) \end{bmatrix}. \tag{3.103}$$

The equilibrium equations for the x-y (see Eqs. (3.28) and (3.27)) and for the x-z plane (see Eqs. (3.76) and (3.75)) can be summarized in matrix notation as follows:

$$\begin{bmatrix} \dfrac{dM_y(x)}{dx} \\ \dfrac{dM_z(x)}{dx} \\ \dfrac{dQ_y(x)}{dx} \\ \dfrac{dQ_z(x)}{dx} \end{bmatrix} = \begin{bmatrix} Q_z(x) \\ -Q_y(x) \\ 0 \\ 0 \end{bmatrix} - \begin{bmatrix} m_y(x) \\ m_z(x) \\ q_y(x) \\ q_z(x) \end{bmatrix}. \tag{3.104}$$

Introducing in Eq. (3.104) the formulation of (3.103), one can obtain the following matrix equation:

$$\frac{d}{dx} \left(\begin{bmatrix} EI_y & EI_{yz} & 0 & 0 \\ EI_{yz} & EI_z & 0 & 0 \\ 0 & 0 & k_s A G & 0 \\ 0 & 0 & 0 & k_s A G \end{bmatrix} \begin{bmatrix} \dfrac{d\phi_y(x)}{dx} \\ \dfrac{d\phi_z(x)}{dx} \\ \dfrac{du_y(x)}{dx} - \phi_z(x) \\ \dfrac{du_z(x)}{dx} + \phi_y(x) \end{bmatrix} \right) = \begin{bmatrix} Q_z(x) \\ -Q_y(x) \\ 0 \\ 0 \end{bmatrix} - \begin{bmatrix} m_y(x) \\ m_z(x) \\ q_y(x) \\ q_z(x) \end{bmatrix}.$$

$$\tag{3.105}$$

If we express the column matrix of the shear forces in Eq. (3.105) based on the relations of Eqs. (3.101) and (3.102), i.e.,

$$
\begin{bmatrix} 0 & 0 & 0 & k_sAG \\ 0 & 0 & k_sAG & 0 \\ 0 & 0 & 0 & 0 \\ 0 & 0 & 0 & 0 \end{bmatrix}
\begin{bmatrix} \dfrac{d\phi_y(x)}{dx} \\[2ex] \dfrac{d\phi_z(x)}{dx} \\[2ex] \dfrac{du_y(x)}{dx} - \phi_z(x) \\[2ex] \dfrac{du_z(x)}{dx} + \phi_y(x) \end{bmatrix} , \tag{3.106}
$$

the system of coupled ordinary differential equations for the determination of $u_y(x)$, $u_z(x)$, $\phi_y(x)$, and $\phi_z(x)$ can be finally expressed as:

$$
\left(\frac{d}{dx} \left[\begin{bmatrix} EI_y & EI_{yz} & 0 & 0 \\ EI_{yz} & EI_z & 0 & 0 \\ 0 & 0 & k_sAG & 0 \\ 0 & 0 & 0 & k_sAG \end{bmatrix}
\begin{bmatrix} \dfrac{d\phi_y(x)}{dx} \\[2ex] \dfrac{d\phi_z(x)}{dx} \\[2ex] \dfrac{du_y(x)}{dx} - \phi_z(x) \\[2ex] \dfrac{du_z(x)}{dx} + \phi_y(x) \end{bmatrix} \right] \right) =
$$

$$
\begin{bmatrix} 0 & 0 & 0 & k_sAG \\ 0 & 0 & k_sAG & 0 \\ 0 & 0 & 0 & 0 \\ 0 & 0 & 0 & 0 \end{bmatrix}
\begin{bmatrix} \dfrac{d\phi_y(x)}{dx} \\[2ex] \dfrac{d\phi_z(x)}{dx} \\[2ex] \dfrac{du_y(x)}{dx} - \phi_z(x) \\[2ex] \dfrac{du_z(x)}{dx} + \phi_y(x) \end{bmatrix}
- \begin{bmatrix} m_y(x) \\ m_z(x) \\ q_y(x) \\ q_z(x) \end{bmatrix} . \tag{3.107}
$$

3.6 Supplementary Problems

3.1 Cantilever Timoshenko beam with different end loads and deformations

Calculate the analytical solution for the deflection $u_z(x)$ and rotation $\phi_y(x)$ of the cantilever Timoshenko beams shown in Fig. 3.17. It can be assumed for this exercise that the bending EI_y and the shear k_sAG stiffnesses are constant. Calculate in addition for all four cases the reactions at the fixed support and the distributions of the bending moment and shear force.

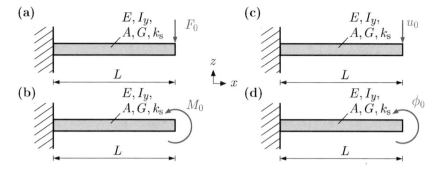

Fig. 3.17 Cantilever Timoshenko beam with different end loads and deformations: **a** single force; **b** single moment; **c** displacement; **d** rotation

3.2 Calculation of the shear correction factor for a rectangular cross section

For a rectangular cross section with width b and height h, the shear stress distribution is given as follows:

$$\tau_{zx}(z) = \frac{6Q_z}{bh^3}\left(\frac{h^2}{4} - z^2\right) \quad \text{with} \quad -\frac{h}{2} \leq z \leq \frac{h}{2}. \tag{3.108}$$

Compute the shear correction factor k_s under the assumption that the constant—in the surface A_s acting—equivalent shear stress $\tau_{xz} = Q_z/A_s$ yields the same shear strain energy as the actual shear stress distribution $\tau_{xz}(z)$, which acts in the actual cross-sectional area A of the beam.

References

1. Altenbach H, Öchsner A (eds) (2020) Encyclopedia of continuum mechanics. Springer, Berlin
2. Bathe K-J (1996) Finite element procedures. Prentice-Hall, Upper Saddle River
3. Beer FP, Johnston ER Jr, DeWolf JT, Mazurek DF (2009) Mechanics of materials. McGraw-Hill, New York
4. Cowper GR (1966) The shear coefficient in Timoshenko's beam theory. J Appl Mech 33:335–340
5. Gere JM, Timoshenko SP (1991) Mechanics of materials. PWS-KENT Publishing Company, Boston
6. Gruttmann F, Wagner W (2001) Shear correction factors in Timoshenko's beam theory for arbitrary shaped cross-sections. Comput Mech 27:199–207
7. Levinson M (1981) A new rectangular beam theory. J Sound Vib 74:81–87
8. Öchsner A (2014) Elasto-plasticity of frame structure elements: modeling and simulation of rods and beams. Springer, Berlin
9. Öchsner A, Merkel M (2018) One-dimensional finite elements: an introduction to the FE method. Springer, Cham

10. Öchsner A (2020) Computational statics and dynamics: an introduction based on the finite element method. Springer, Singapore
11. Reddy JN (1984) A simple higher-order theory for laminated composite plate. J Appl Mech 51:745–752
12. Reddy JN (1997) Mechanics of laminated composite plates: theory and analysis. CRC Press, Boca Raton
13. Reddy JN (1997) On locking-free shear deformable beam finite elements. Comput Method Appl M 149:113–132
14. Timoshenko SP (1921) On the correction for shear of the differential equation for transverse vibrations of prismatic bars. Philos Mag 41:744–746
15. Timoshenko SP (1922) On the transverse vibrations of bars of uniform cross-section. Philos Mag 43:125–131
16. Timoshenko S (1940) Strength of materials - part I elementary theory and problems. D. Van Nostrand Company, New York
17. Timoshenko SP, Goodier JN (1970) Theory of elasticity. McGraw-Hill, New York
18. Twiss RJ, Moores EM (1992) Structural geology. WH Freeman & Co, New York
19. Weaver W Jr, Gere JM (1980) Matrix analysis of framed structures. Van Nostrand Reinhold Company, New York
20. Winkler E (1867) Die Lehre von der Elasticität und Festigkeit mit besonderer Rücksicht auf ihre Anwendung in der Technik. H. Dominicus, Prag

Chapter 4
Higher-Order Beam Theories

Abstract This chapter presents the analytical description of thick, or so-called shear-flexible, beam members according to the Levinson theory. Based on the three basic equations of continuum mechanics, i.e., the kinematics relationship, the constitutive law, and the equilibrium equation, the partial differential equations, which describe the physical problem, are presented. All equations are introduced for single plane bending in the x-y plane as well as the x-z plane. Analytical solutions of the partial differential equations are given for simple cases. In addition, this chapter treats the case of unsymmetrical bending and presents a general systematic for higher-order beam theories.

4.1 Introduction

The different beam theories can be introduced in different ways. One discrimination is the different consideration of the shear stress to derive the describing differential equations of the deformation, see Fig. 4.1. The Euler–Bernoulli beam theory (see Fig. 4.1a), which was introduced in Chap. 2, neglects the influence of the shear stress on the deformed shape. The Timoshenko beam theory (see Fig. 4.1b) assumes an equivalent *constant* shear stress in the entire cross section as introduced in Chap. 3. Higher-order beam theories consider a more realistic distribution of the shear stress (see Fig. 4.1c). A good introduction to this topic can be found in the monograph [11], the proceedings book [2] and the review article [3].

Systematics of the different theories can be given based on the assumed displacement fields $u_x(x, y)$ and $u_y(x, y)$ (or $u_x(x, z)$ and $u_z(x, z)$) [3, 11]. This implies a different deformation of a typical transverse normal line which is in the undeformed state perpendicular to the center line (mid-plane), see Figs. 4.2 and 4.3.

- The Euler–Bernoulli or elementary beam theory (EBT) assumes a displacement field in the x-y plane of the form:

$$u_x(x, y) = -y\frac{\mathrm{d}u_y(x)}{\mathrm{d}x} = -y\varphi_z ,\tag{4.1}$$

$$u_y(x, y) = u_y(x) ,\tag{4.2}$$

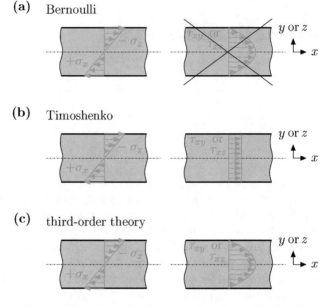

Fig. 4.1 Comparison of different beam theories in regards to the contribution of the stress state on the deformation: **a** Euler–Bernoulli, **b** Timoshenko and **c** higher-order theory. Rectangular cross section assumed

or in the x-z plane:

$$u_x(x, z) = -z\frac{\mathrm{d}u_z(x)}{\mathrm{d}x} = +z\varphi_y \,, \tag{4.3}$$

$$u_z(x, z) = u_z(x)\,. \tag{4.4}$$

This implies that the plane sections which are perpendicular to the neutral fiber before bending remain plane and perpendicular to the neutral fiber after bending, see Figs. 4.2a and 4.3a.

• The Timoshenko beam or first-order shear deformation theory (FSDT) assumes a displacement field in the x-y plane of the form:

$$u_x(x, y) = -y\phi_z \,, \tag{4.5}$$

$$u_y(x, y) = u_y(x)\,, \tag{4.6}$$

or in the x-z plane:

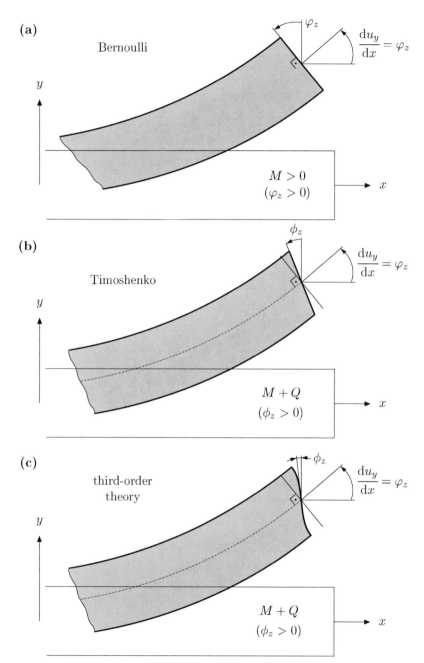

Fig. 4.2 Comparison of different beam theories in regards to the deformation in the x-y plane of a typical transverse normal plane: **a** Euler–Bernoulli, **b** Timoshenko and **c** higher-order theory. Note that the deformation is overdrawn for better illustration

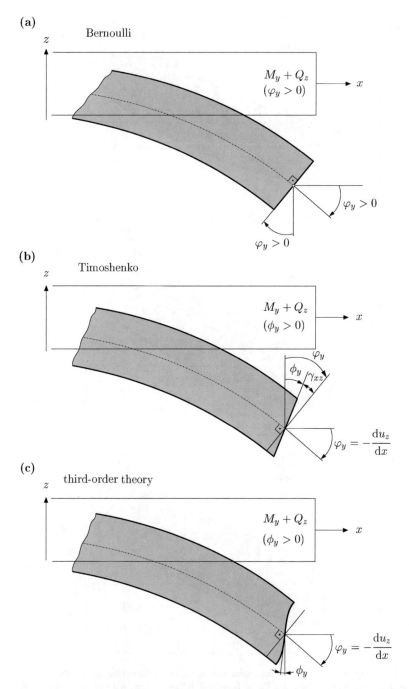

Fig. 4.3 Comparison of different beam theories in regards to the deformation in the x-z plane of a typical transverse normal plane: **a** Euler–Bernoulli, **b** Timoshenko and **c** higher-order theory. Note that the deformation is overdrawn for better illustration

$$u_x(x, z) = +z\phi_y \,, \tag{4.7}$$

$$u_z(x, z) = u_z(x) \,. \tag{4.8}$$

This implies that the plane sections which are perpendicular to the neutral fiber before bending are no longer perpendicular to the neutral fiber after bending but they remain plane, see Figs. 4.2b and 4.3b.

• A second-order shear deformation theory (SSDT) assumes a displacement field in the x-y plane of the form:

$$u_x(x, y) = -y\phi_z(x) + y^2\psi(x) \,, \tag{4.9}$$

$$u_y(x, y) = u_y(x) \,, \tag{4.10}$$

or in the x-z plane:

$$u_x(x, z) = +z\phi_y(x) + z^2\psi(x) \,, \tag{4.11}$$

$$u_z(x, z) = u_z(x) \,. \tag{4.12}$$

Now, an originally plane section is, after deformation, no longer perpendicular to the neutral fiber and no longer plane. The functions $\phi_z(x)$ or $\phi_y(x)$ represent the rotation of the cross section of the beam at neutral fiber ($y = 0$ or $z = 0$) and $\phi_z(x)$ together with $\psi(x)$ defines the quadratic nature of the deformed line.

• A third-order shear deformation theory (TSDT) assumes a displacement field in the x-y plane of the form:

$$u_x(x, y) = -y\phi_z(x) + y^2\psi(x) + y^3\theta(x) \,, \tag{4.13}$$

$$u_y(x, y) = u_y(x) \,, \tag{4.14}$$

or in the x-z plane:

$$u_x(x, z) = +z\phi_y(x) + z^2\psi(x) + z^3\theta(x) \,, \tag{4.15}$$

$$u_z(x, z) = u_z(x) \,. \tag{4.16}$$

Here, $\phi_z(x)$ and $\phi_y(x)$ represent again the rotation of the cross section of the beam at neutral fiber ($y = 0$ or $z = 0$), see Figs. 4.2c and 4.3c.

It can be seen from the above that the second- and third-order theories are expressed as an extension of the Timoshenko theory. In a slightly different form of systematics [9, 10], the higher-order theories may be expressed as extensions of the Euler–Bernoulli beam theory in the x-y plane as:

$$u_x(x, y) = -y\frac{du_y(x)}{dx} + f(y)\phi_z(x),\qquad(4.17)$$

$$u_y(x, y) = u_y(x),\qquad(4.18)$$

or in the x-z plane:

$$u_x(x, z) = -z\frac{du_z(x)}{dx} + f(z)\phi_y(x),\qquad(4.19)$$

$$u_z(x, z) = u_z(x).\qquad(4.20)$$

Let us consider in the following a classical example of third-order theory which was proposed by Levinson in [4]. His proposed displacement field is generally expressed in the x-y plane as:

$$u_x(x, y) = -y\phi_z(x) + y^3\theta(x),\qquad(4.21)$$

$$u_y(x, y) = u_y(x),\qquad(4.22)$$

or in the x-z plane:

$$u_x(x, z) = +z\phi_y(x) + z^3\theta(x),\qquad(4.23)$$

$$u_z(x, z) = u_z(x).\qquad(4.24)$$

4.2   Deformation in the x-y Plane

4.2.1   Kinematics

With the general relationship for the normal strain, i.e., $\varepsilon_x(x, y) = \partial u_x(x, y)/\partial x$, one obtains the kinematics relation from the displacement field (4.21) as:

$$\varepsilon_x(x, y) = -y\frac{d\phi_z(x)}{dx} + y^3\frac{d\theta(x)}{dx}.\qquad(4.25)$$

It should be noted here that the first expression in Eq. (4.25) represents the relationship for the Timoshenko beam according to Eq. (3.17). Furthermore, the requirement that the shear stress must vanish at the outer surface layers (see Fig. 4.1c), i.e., at $y = \pm h/2$, (rectangular cross section $-h/2 \le y \le h/2$ and $-b/2 \le z \le b/2$ assumed) yields:

$$\varepsilon_{xy}(x, \pm\tfrac{h}{2}) = \frac{1}{2}\left(\frac{\partial u_y(x)}{\partial x} + \frac{\partial u_x(x, y)}{\partial y}\right)\bigg|_{y=\pm\frac{h}{2}} \overset{!}{=} 0. \qquad (4.26)$$

Under the assumption of a rectangular cross section ($b \times h$) and the relationship

$$\frac{\partial u_x(x, y)}{\partial z}\bigg|_{y=\pm\frac{h}{2}} = -\phi_z(x) + 3y^2\theta(x)\big|_{y=\pm\frac{h}{2}} = -\phi_z(x) + \frac{3h^2}{4}\theta(x), \qquad (4.27)$$

the above stress condition (4.26) can be rearranged for $\theta(x)$ to obtain:

$$\theta(x) = -\frac{4}{3h^2}\left(-\phi_z(x) + \frac{\partial u_y(x)}{\partial x}\right). \qquad (4.28)$$

The last relationship can be used in the displacement field according to (4.21) and the shear strain according to Eq. (3.10) is obtained as:

$$\gamma_{xy}(x, y) = \frac{h^2 - 4y^2}{h^2}\left(-\phi_z(x) + \frac{\partial u_y(x)}{\partial x}\right). \qquad (4.29)$$

It should be noted here that the last equation simplifies for $y = 0$ to the Timoshenko relationship according to Eq. (3.18).

4.2.2 Constitutive Equation

For the consideration of the constitutive relation, Hooke's law for a one-dimensional normal stress state and for a one-dimensional shear stress state is again used (see Sect. 3.2.2):

$$\sigma_x(x, y) = E\varepsilon_x(x, y), \qquad (4.30)$$
$$\tau_{xy}(x, y) = G\gamma_{xy}(x, y). \qquad (4.31)$$

According to the equilibrium configuration of Fig. 2.9 and Eq. (2.22), the relation between the internal bending moment and the bending stress can be used for the Levinson beam as follows: Consideration of the constitutive relation (4.30) and the kinematics relation $\varepsilon_x(x, y) = \partial u_x(x, y)/\partial x$ gives:

$$M_z(x) = -\int_A y\sigma_x dA = -\int_{z=-\frac{b}{2}}^{+\frac{b}{2}} \int_{y=-\frac{h}{2}}^{+\frac{h}{2}} yE\frac{\partial u_x(x,y)}{\partial x} dzdy. \qquad (4.32)$$

Based on the general expression for the displacement field (4.21) and the function $\theta(x)$ given in Eq. (4.28), the partial derivative of the displacement field with respect to the x-coordinate can be expressed as:

$$\frac{\partial u_x(x,y)}{\partial x} = \frac{-3h^2 y + 4y^3}{3h^2}\frac{\partial\phi_z(x)}{\partial x} - \frac{4y^3}{3h^2}\frac{\partial^2 u_y(x)}{\partial x^2}. \qquad (4.33)$$

Introducing this last expression in Eq. (4.32) and assuming a constant stiffness E as well as a rectangular cross section ($b \times h$), one obtains:

$$M_z(x) = \frac{EI_z}{5}\left(4\frac{\partial\phi_z(x)}{\partial x} + \frac{\partial^2 u_y(x)}{\partial x^2}\right). \qquad (4.34)$$

The internal shear force results from Eq. (3.11) and the application of the constitutive relation (4.31) as:

$$Q_y(x) = \int_A \tau_{xy}(x,y)dA = \int_{z=-\frac{b}{2}}^{+\frac{b}{2}} \int_{y=-\frac{h}{2}}^{+\frac{h}{2}} \gamma_{xy}(x,y)Gdzdy. \qquad (4.35)$$

Consideration of the relation for the shear strain according to Eq. (3.10) and assuming a constant shear modulus G as well as a rectangular cross section ($b \times h$), one obtains:

$$Q_y(x) = \frac{2}{3}GA\left(-\phi_z(x) + \frac{\partial u_y(x)}{\partial x}\right). \qquad (4.36)$$

Similar to Sects. 2.2.2 and 3.2.2, one can use the stress resultants to formulate the constitutive law based on the generalized stresses $s = \begin{bmatrix} M_z, & Q_y \end{bmatrix}^T$ and generalized strains $e = \left[\frac{1}{5}\left(4\frac{\partial\phi_z(x)}{\partial x} + \frac{\partial^2 u_y(x)}{\partial x^2}\right), \frac{2}{3}\left(-\phi_z + \frac{du_y}{dx}\right)\right]^T$, see Fig. 4.4. The generalized quantities reveal the advantage that they are not a function of the vertical coordinate.

4.2.3 Equilibrium

The derivation of the equilibrium condition for the Levinson beam is identical with the derivation for the Euler–Bernoulli beam according to Sect. 2.2.3:

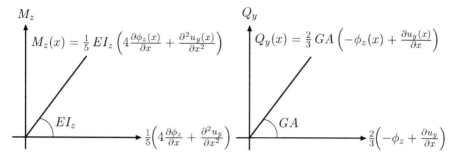

Fig. 4.4 Formulation of the constitutive law based on **a** generalized normal stress and **b** generalized shear stress (bending in the x-y plane)

Table 4.1 Elementary basic equations for the bending of a Levinson beam in the x-y plane (y-axis: right facing; z-axis: upward facing)

Relation	Equation
Kinematics	$\varepsilon_x(x, y) = -y \frac{d\phi_z(x)}{dx} + y^3 \frac{d\theta(x)}{dx}$ and
	$\gamma_{xy}(x, y) = \frac{h^2 - 4y^2}{h^2}\left(-\phi_z(x) + \frac{\partial u_y(x)}{\partial x}\right)$
Equilibrium	$\dfrac{dQ_y(x)}{dx} = -q_y(x)$
	$\dfrac{dM_z(x)}{dx} = -Q_y(x) - m_z(x)$
Constitution	$\sigma_x(x, y) = E\varepsilon_x(x, y)$ and
	$\tau_{xy}(x, y) = G\gamma_{xy}(x, y)$

$$\frac{dQ_y(x)}{dx} = -q_y(x)\,, \tag{4.37}$$

$$\frac{dM_z(x)}{dx} = -Q_y(x) - m_z(x)\,. \tag{4.38}$$

Before looking in more detail at the differential equations of the bending line, let us summarize the basic equations for the Levinson beam in Table 4.1. It should be noted that the shear stress and shear strain are now dependent on the x- and z-coordinate.

4.2.4 Differential Equation

Using the expressions for the internal bending moment (4.34) and the internal shear force (4.36) in the equilibrium conditions (4.37) and (4.38) gives finally the coupled systems of partial differential equations for the Levinson beam as:

$$\frac{2}{3}\frac{\partial}{\partial x}\left[GA\left(-\phi_z(x)+\frac{\partial u_y(x)}{\partial x}\right)\right]=-q_y(x),$$

$$(4.39)$$

$$\frac{1}{5}\frac{\partial}{\partial x}\left[EI_z\left(4\frac{\partial\phi_z(x)}{\partial x}+\frac{\partial^2 u_y(x)}{\partial x^2}\right)\right]+\frac{2}{3}GA\left(-\phi_z(x)+\frac{\partial u_y(x)}{\partial x}\right)=-m_z(x).$$

$$(4.40)$$

This systems contains—as in the case of the Timoshenko beam—two unknown functions, i.e. the deflection $u_y(x)$ and the cross-sectional rotation $\phi_z(x)$ at $y=0$.

Different formulations of these coupled differential equations are collected in Table 4.2 where different types of loadings, geometry and bedding are differentiated. The last case in Table 4.2 refers again to the elastic or Winkler foundation of a beam, [12]. The elastic foundation modulus k has in the case of beams the unit of force per unit area.

A single-equation description for the Levinson beam can be obtained under the assumption of constant material (E, G) and geometrical (I_z, A) properties: Rearranging and two-times differentiation of Eq. (4.39) gives:

$$\frac{d\phi_z(x)}{dx}=\frac{d^2 u_y(x)}{dx^2}+\frac{3q_y(x)}{2GA},$$

$$(4.41)$$

$$\frac{d^3\phi_z(x)}{dx^3}=\frac{d^4 u_y(x)}{dx^4}+\frac{3d^2 q_y(x)}{2GAdx^2}.$$

$$(4.42)$$

One-time differentiation of Eq. (4.40) gives:

$$\frac{1}{5}EI_z\left(4\frac{d^3\phi_z(x)}{dx^3}+\frac{d^4 u_y(x)}{dx^4}\right)+\frac{2}{3}AG\left(\frac{d^2 u_y(x)}{dx^2}-\frac{d\phi_z(x)}{dx}\right)=-\frac{dm_z(x)}{dx}.$$

$$(4.43)$$

Inserting Eq. (4.42) into (4.43) and consideration of (4.41) gives finally the following expression:

$$EI_z\frac{d^4 u_y(x)}{dx^4}=q_y(x)-\frac{dm_z(x)}{dx}-\frac{EI_z}{\frac{5}{6}AG}\frac{d^2 q_y(x)}{dx^2}.$$

$$(4.44)$$

The last equation reduces for shear-rigid beams, i.e. $\frac{5}{6}AG\to\infty$, to the classical Euler–Bernoulli formulation as given in Table 2.4.

If we replace in the previous formulations the first- and second-order derivatives, i.e. $\frac{d...}{dx}$ and $\frac{d^2...}{dx^2}$, by formal operator symbols, i.e. the $\mathcal{L}_1$ and $\mathcal{L}_2$–matrices, then the basic equations of the Levinson beam can be stated in a more formal way as given in Table 4.3. Such a matrix formulation is more suitable for the derivation of the principal finite element equation based on the weighted residual method [7, 8].

Table 4.2 Different formulations of the partial differential equation for a Levinson beam in the x-y plane (x-axis: right facing; y-axis: upward facing)

Configuration	Partial differential equation
$\boxed{E, I_y, A, G}$	$\dfrac{EI_z}{5}\left(4\dfrac{\partial^2\phi_z(x)}{\partial x^2} + \dfrac{\partial^3 u_y(x)}{\partial x^3}\right) +$ $\dfrac{2}{3}GA\left(-\phi_z(x) + \dfrac{\partial u_y(x)}{\partial x}\right) = 0$ $\dfrac{2GA}{3}\left(-\dfrac{\partial\phi_z(x)}{\partial x} + \dfrac{\partial^2 u_y(x)}{\partial x^2}\right) = 0$
$E(x), I_y(x)$ $A(x), G(x)$	$\dfrac{1}{5}\dfrac{\partial}{\partial x}\left[EI_z\left(4\dfrac{\partial\phi_z(x)}{\partial x} + \dfrac{\partial^2 u_y(x)}{\partial x^2}\right)\right] +$ $\dfrac{2}{3}GA\left(-\phi_z(x) + \dfrac{\partial u_y(x)}{\partial x}\right) = 0$ $\dfrac{2}{3}\dfrac{\partial}{\partial x}\left[GA\left(-\phi_z(x) + \dfrac{\partial u_y(x)}{\partial x}\right)\right] = 0$
$q_z(x)$	$\dfrac{EI_z}{5}\left(4\dfrac{\partial^2\phi_z(x)}{\partial x^2} + \dfrac{\partial^3 u_y(x)}{\partial x^3}\right) +$ $\dfrac{2}{3}GA\left(-\phi_z(x) + \dfrac{\partial u_y(x)}{\partial x}\right) = 0$ $\dfrac{2GA}{3}\left(-\dfrac{\partial\phi_z(x)}{\partial x} + \dfrac{\partial^2 u_y(x)}{\partial x^2}\right) = -q_y(x)$
$m_y(x)$	$\dfrac{EI_z}{5}\left(4\dfrac{\partial^2\phi_z(x)}{\partial x^2} + \dfrac{\partial^3 u_y(x)}{\partial x^3}\right) +$ $\dfrac{2}{3}GA\left(-\phi_z(x) + \dfrac{\partial u_y(x)}{\partial x}\right) = -m_z(x)$ $\dfrac{2GA}{3}\left(-\dfrac{\partial\phi_z(x)}{\partial x} + \dfrac{\partial^2 u_y(x)}{\partial x^2}\right) = 0$
$k(x)$	$\dfrac{EI_z}{5}\left(4\dfrac{\partial^2\phi_z(x)}{\partial x^2} + \dfrac{\partial^3 u_y(x)}{\partial x^3}\right) +$ $\dfrac{2}{3}GA\left(-\phi_z(x) + \dfrac{\partial u_y(x)}{\partial x}\right) = 0$ $\dfrac{2GA}{3}\left(-\dfrac{\partial\phi_z(x)}{\partial x} + \dfrac{\partial^2 u_y(x)}{\partial x^2}\right) = k(x)u_y(x)$

Table 4.3 Different formulations of the basic equations for a Levinson beam (bending in the x-y plane; x-axis along the principal beam axis). E: Young's modulus; G: shear modulus; A: cross-sectional area; I_z: second moment of area; q_y: length-specific distributed force; m_z: length-specific distributed moment; e: generalized strains; s^*: generalized stresses

Specific formulation	General formulation [1]
Kinematics	
$\begin{bmatrix} \frac{du_y}{dx} - \phi_z \\ \frac{d^2 u_y}{dx^2} + 4\frac{d\phi_z}{dx} \end{bmatrix} = \begin{bmatrix} \frac{d}{dx} & -1 \\ \frac{d^2}{dx^2} & \frac{d}{dx} \end{bmatrix} \begin{bmatrix} u_y \\ \phi_z \end{bmatrix}$	$e = \mathcal{L}_{2*} u$
Constitution	
$\begin{bmatrix} Q_y \\ M_z \end{bmatrix} = \begin{bmatrix} \frac{2}{3} AG & 0 \\ 0 & \frac{1}{5} E I_z \end{bmatrix} \begin{bmatrix} \frac{du_y}{dx} - \phi_z \\ \frac{d^2 u_y}{dx^2} + 4\frac{d\phi_z}{dx} \end{bmatrix}$	$s^* = D^* e$
Equilibrium	
$\begin{bmatrix} \frac{d}{dx} & 0 \\ 1 & \frac{d}{dx} \end{bmatrix} \begin{bmatrix} Q_y \\ M_z \end{bmatrix} + \begin{bmatrix} q_y \\ m_z \end{bmatrix} = \begin{bmatrix} 0 \\ 0 \end{bmatrix}$	$\mathcal{L}_1^T s^* + b^* = 0$
PDE	
$\dfrac{d}{dx}\left[\dfrac{2}{3} GA \left(\dfrac{du_y}{dx} - \phi_z \right) \right] + q_y = 0$ $\dfrac{d}{dx}\left[\dfrac{1}{5} E I_z \left(\dfrac{d^2 u_y}{dx^2} + \dfrac{d\phi_z}{dx} \right) \right] +$ $+ \dfrac{2}{3} GA \left(\dfrac{du_y}{dx} - \phi_z \right) + m_z = 0 ,$	$\mathcal{L}_1^T D^* \mathcal{L}_{2*} u + b^* = 0$

Under the assumption of constant material (E, G) and geometric (I_z, A) properties, the system of differential equations in Table 4.3 can be solved by using a computer algebra system for constant distributed loads $(q_y(x) = q_0 = \text{const.}$ and $m_z(x) = m_0 = \text{const.})$ to obtain the general analytical solution of the problem, see Fig. 4.5.

The general analytical solution provided in Fig. 4.5, i.e.,

$$u_y(x) = - \frac{\begin{aligned} &36 AG\, E I_z x^2 \left(\frac{d^2 u_y(0)}{dx^2} \right) + \left(360 AG\, E I_z x - 40(AG)^2 x^3 \right) \left(\frac{du_y(0)}{dx} \right) \\ &+ 144 AG\, E I_z x^2 \left(\frac{d\phi_z(0)}{dx} \right) + 15 AG q_0 x^4 \\ &+ \left(-60 AGm + 40\phi_z(0)(AG)^2 \right) x^3 - 216 E I_z q_0 x^2 + 360 u_y(0) AG\, E I_z \end{aligned}}{360 AG E I_z} ,$$

$$\phi_z(x) = \frac{\begin{aligned} &6 AG\, E I_z x \left(\frac{d^2 u_y(0)}{dx^2} \right) - 10(AG)^2 x^2 \left(\frac{du_y(0)}{dx} \right) + 24 AG\, E I_z x \left(\frac{d\phi_z(0)}{dx} \right) \\ &+ 5 AG q_0 x^3 + \left(-15 AG m_0 + 10\phi_z(0)(AG)^2 \right) x^2 \\ &+ 9 E I_z q_0 x + 30\phi_z(0) AG\, E I_z \end{aligned}}{30 AG E I_z} ,$$

Solution of the coupled DEs for a Levinson beam with AG, EI, q, m = const

```
(%i14)   eqn_1: 2/3*AG*'diff(u(x),x,2) = -q+2/3*AG*'diff(phi(x),x)$
         eqn_2: 1/5*EI*'diff(u(x),x,3)+2/3*AG*'diff(u(x),x) = -m-4/5*EI*'diff(phi(x),x,2)
                                                              +2/3*AG*phi(x)$

         sol : desolve([eqn_1, eqn_2], [u(x), phi(x)])$

         print(" ")$
         print("Equations:")$
         print(" ", eqn_1)$
         print(" ")$
         print(" ", eqn_2)$
         print(" ")$
         print(" ")$
         print("Analytical Solution:")$
         print(ratsimp(sol[1]))$
         print(" ")$
         print(ratsimp(sol[2]))$
```

Equations:

$$\frac{2}{3} AG \left(\frac{d^2}{dx^2} u(x) \right) = \frac{2}{3} AG \left(\frac{d}{dx} \phi(x) \right) - q$$

$$\frac{1}{5} EI \left(\frac{d^3}{dx^3} u(x) \right) + \frac{2}{3} AG \left(\frac{d}{dx} u(x) \right) = -\frac{4}{5} EI \left(\frac{d^2}{dx^2} \phi(x) \right) + \frac{2}{3} AG\, \phi(x) - m$$

Analytical Solution:

$$u(x) = -\frac{\begin{aligned} &36 AG\,EIx^2 \left(\frac{d^2}{dx^2} u(x) \big|_{x=0} \right) + (360 AG\,EIx - 40 AG^2 x^3) \left(\frac{d}{dd} u(x) \big|_{x=0} \right) \\ &+ 144 AG\,EIx^2 \left(\frac{d}{dx} phi(x) \big|_{x=0} \right) + 15 AGqx^4 \\ &+ (-60 AGm + 40\, phi(0)AG^2) x^3 - 216 EIqx^2 + 360\, u(0)AG\,EI \end{aligned}}{360 AGEI}$$

$$phi(x) = \frac{\begin{aligned} &6 AG\,EIx \left(\frac{d^2}{dx^2} u(x) \big|_{x=0} \right) - 10 AG^2 x^2 \left(\frac{d}{dx} u(x) \big|_{x=0} \right) \\ &+ 24 AG\,EIx \left(\frac{d}{dx} phi(x) \big|_{x=0} \right) + 5 AGqx^3 \\ &+ (-15 AGm + 10\, phi(0)AG^2) x^2 + 9 EIqx + 30\, phi(0)AG\,EI \end{aligned}}{30 AGEI}$$

Fig. 4.5 Solution of the coupled system of differential equations for the Levinson beams based on the computer algebra system Maxima

where $\frac{\mathrm{d}u_y(0)}{\mathrm{d}x}$, $\frac{\mathrm{d}\phi_z(0)}{\mathrm{d}x}$, $u_y(0)$, and $\phi_z(0)$ are the four boundary values to adjust the general solution to a particular problem, can be represented in a slightly different way based on a different set of constants:

$$
u_y(x) = \frac{1}{EI_z}\left(\frac{q_0 x^4}{24} + \frac{x^3}{6}\underbrace{\left[-\frac{2}{3}AG\frac{\mathrm{d}u_y(0)}{\mathrm{d}x} + \frac{2}{3}AG\phi_z(0)\right]}_{c_1} + \right.
$$

$$
+ \frac{x^2}{2}\underbrace{\left[\frac{EI_z}{5}\frac{\mathrm{d}^2 u_y(0)}{\mathrm{d}x^2} + \frac{4EI_z}{5}\frac{\mathrm{d}\phi_z(0)}{\mathrm{d}x} - \frac{6EI_z q_0}{5 AG}\right]}_{c_2} + \underbrace{EI_z\frac{\mathrm{d}u_y(0)}{\mathrm{d}x}}_{c_3}x +
$$

$$
\left. + \underbrace{EI_z u_y(0)}_{c_4}\right) - \frac{m_0 x^3}{6EI_z}, \tag{4.45}
$$

$$
\phi_z(x) = \frac{1}{EI_z}\left(\frac{q_0 x^3}{3} + \frac{x^2}{2}\underbrace{\left[-\frac{2}{3}AG\frac{\mathrm{d}u_y(0)}{\mathrm{d}x} + \frac{2}{3}AG\phi_z(0)\right]}_{c_1} + \right.
$$

$$
\left. + x\underbrace{\left[\frac{EI_z}{5}\frac{\mathrm{d}^2 u_y(0)}{\mathrm{d}x^2} + \frac{4EI_z}{5}\frac{\mathrm{d}\phi_z(0)}{\mathrm{d}x} - \frac{6EI_z q_0}{5 AG}\right]}_{c_2} + \underbrace{EI_z\frac{\mathrm{d}u_y(0)}{\mathrm{d}x}}_{c_3}\right)
$$

$$
+ \frac{q_0 x}{\frac{2}{3}AG} + \underbrace{\left[-\frac{\mathrm{d}u_y(0)}{\mathrm{d}x} + \phi_z(0)\right]}_{\frac{+c_1}{\frac{2}{3}AG}} - \frac{m_0 x^2}{2EI_z}. \tag{4.46}
$$

The general expressions for the internal bending moment and shear force distributions can be obtained under consideration of Eqs. (4.34) and (4.36). Thus, we can finally state the general solution for constant material and geometrical properties as well as constant distributed loads as [5]:

$$u_y(x) = \frac{1}{EI_z}\left(\frac{q_0 x^4}{24} + c_1\frac{x^3}{6} + c_2\frac{x^2}{2} + c_3 x + c_4\right) - \frac{m_0 x^3}{6EI_z}, \tag{4.47}$$

$$\phi_z(x) = \frac{1}{EI_y}\left(\frac{q_0 x^3}{6} + c_1\frac{x^2}{2} + c_2 x + c_3\right) + \frac{q_0 x}{\frac{2}{3}AG} + \frac{c_1}{\frac{2}{3}AG} - \frac{m_0 x^2}{2EI_z}, \tag{4.48}$$

$$M_y(x) = +\left(\frac{q_0 x^2}{2} + c_1 x + c_2\right) + \frac{\frac{4}{5}q_0 EI_y}{\frac{2}{3}AG} - m_0 x, \tag{4.49}$$

$$Q_z(x) = -(q_0 x + c_1), \tag{4.50}$$

where the four constants of integration c_i ($i = 1, \ldots, 4$) must be determined based on the boundary conditions.

4.3 Deformation in the *x*-*z* Plane

4.3.1 Kinematics

With the general relationship for the normal strain, i.e., $\varepsilon_x(x, z) = \partial u_x(x, z)/\partial x$, one obtains the kinematics relation from the displacement field (4.21) as:

$$\varepsilon_x(x, z) = z\frac{d\phi_y(x)}{dx} + z^3\frac{d\theta(x)}{dx}. \tag{4.51}$$

It should be noted here that the first expression in Eq. (4.51) represents the relationship for the Timoshenko beam according to Eq. (3.65). Furthermore, the requirement that the shear stress must vanish at the outer surface layers (see Fig. 4.1c), i.e., at $z = \pm h/2$, (rectangular cross section $-b/2 \le y \le b/2$ and $-h/2 \le z \le h/2$ assumed)[1] yields:

$$\varepsilon_{xz}(x, \pm\tfrac{h}{2}) = \frac{1}{2}\left(\frac{\partial u_z(x)}{\partial x} + \frac{\partial u_x(x, z)}{\partial z}\right)\Bigg|_{z=\pm\frac{h}{2}} \overset{!}{=} 0. \tag{4.52}$$

Under the assumption of a rectangular cross section ($h \times b$) and the relationship

$$\frac{\partial u_x(x, z)}{\partial z}\Bigg|_{z=\pm\frac{h}{2}} = \phi_y(x) + 3z^2\theta(x)\big|_{z=\pm\frac{h}{2}} = \phi_y(x) + \frac{3h^2}{4}\theta(x), \tag{4.53}$$

---

[1]The same rectangular cross section as in Sect. 4.2.2 is assumed here. However, the coordinate system is rotated, i.e. the *z*-axis is parallel to the dimension *h*.

the above stress condition (4.52) can be rearranged for $\theta(x)$ to obtain:

$$\theta(x) = -\frac{4}{3h^2}\left(\phi_y(x) + \frac{\partial u_z(x)}{\partial x}\right).$$

(4.54)

The last relationship can be used in the displacement field according to (4.23) and the shear strain according to Eq. (3.58) is obtained as:

$$\gamma_{xz}(x, z) = \frac{h^2 - 4z^2}{h^2}\left(\phi_y(x) + \frac{\partial u_z(x)}{\partial x}\right).$$

(4.55)

It should be noted here that the last equation simplifies for $z = 0$ to the Timoshenko relationship according to Eq. (3.66).

4.3.2 Constitutive Equation

For the consideration of the constitutive relation, Hooke's law for a one-dimensional normal stress state and for a one-dimensional shear stress state is again used (see Sect. 3.3.2):

$$\sigma_x(x, z) = E\varepsilon_x(x, z),$$

(4.56)

$$\tau_{xz}(x, z) = G\gamma_{xz}(x, z).$$

(4.57)

According to the equilibrium configuration of Fig. 2.19 and Eq. (2.62), the relation between the internal bending moment and the bending stress can be used for the Levinson beam as follows: Consideration of the constitutive relation (4.56) and the kinematics relation $\varepsilon_x(x, z) = \partial u_x(x, z)/\partial x$ gives:

$$M_y(x) = \int_A z\sigma_x dA = \int_{y=-\frac{b}{2}}^{+\frac{b}{2}}\int_{z=-\frac{h}{2}}^{+\frac{h}{2}} zE\frac{\partial u_x(x, z)}{\partial x}\,dy\,dz.$$

(4.58)

Based on the general expression for the displacement field (4.23) and the function $\theta(x)$ given in Eq. (4.54), the partial derivative of the displacement field with respect to the x-coordinate can be expressed as:

$$\frac{\partial u_x(x, z)}{\partial x} = \frac{3h^2 z - 4z^3}{3h^2}\frac{\partial\phi_y(x)}{\partial x} - \frac{4z^3}{3h^2}\frac{\partial^2 u_z(x)}{\partial x^2}.$$

(4.59)

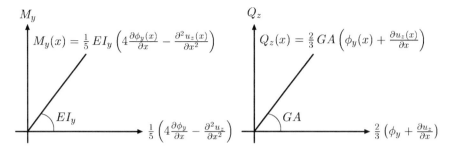

Fig. 4.6 Formulation of the constitutive law based on **a** generalized normal stress and **b** generalized shear stress (bending in the x-z plane)

Introducing this last expression in Eq. (4.58) and assuming a constant stiffness E as well as a rectangular cross section ($b \times h$), one obtains:

$$M_y(x) = \frac{EI_y}{5}\left(4\frac{\partial \phi_y(x)}{\partial x} - \frac{\partial^2 u_z(x)}{\partial x^2}\right).$$ (4.60)

The internal shear force results from Eq. (3.59) and the application of the constitutive relation (4.57) as:

$$Q_z(x) = \int_A \tau_{xz}(x, z)\mathrm{d}A = \int_{y=-\frac{b}{2}}^{+\frac{b}{2}}\int_{z=-\frac{h}{2}}^{+\frac{h}{2}} \gamma_{xz}(x, z)G\mathrm{d}y\mathrm{d}z.$$ (4.61)

Consideration of the relation for the shear strain according to Eq. (3.58) and assuming a constant shear modulus G as well as a rectangular cross section ($b \times h$), one obtains:

$$Q_z(x) = \frac{2}{3}GA\left(\phi_y(x) + \frac{\partial u_z(x)}{\partial x}\right).$$ (4.62)

Similar to Sects. 2.3.2 and 3.3.2, one can use the stress resultants to formulate the constitutive law based on the generalized stresses $s = \begin{bmatrix} M_y, & Q_z \end{bmatrix}^T$ and generalized strains $e = \begin{bmatrix} \frac{1}{5}\left(4\frac{\partial \phi_y(x)}{\partial x} - \frac{\partial^2 u_z(x)}{\partial x^2}\right), & \frac{2}{3}\left(\phi_y + \frac{\mathrm{d}u_z}{\mathrm{d}x}\right) \end{bmatrix}^T$, see Fig. 4.6. The generalized quantities reveal the advantage that they are not a function of the vertical coordinate.

Table 4.4 Elementary basic equations for the bending of a Levinson beam in the x-z plane (x-axis: right facing; z-axis: upward facing)

Relation	Equation
Kinematics	$\varepsilon_x(x, z) = z\frac{\mathrm{d}\phi_y(x)}{\mathrm{d}x} + z^3 \frac{\mathrm{d}\theta(x)}{\mathrm{d}x}$ and $\gamma_{xz}(x, z) = \frac{h^2 - 4z^2}{h^2}\left(\phi_y(x) + \frac{\partial u_z(x)}{\partial x}\right)$
Equilibrium	$\dfrac{\mathrm{d}Q_z(x)}{\mathrm{d}x} = -q_z(x)$ $\dfrac{\mathrm{d}M_y(x)}{\mathrm{d}x} = +Q_z(x) - m_y(x)$
Constitution	$\sigma_x(x, z) = E\varepsilon_x(x, z)$ and $\tau_{xz}(x, z) = G\gamma_{xz}(x, z)$

4.3.3 Equilibrium

The derivation of the equilibrium condition for the Levinson beam is identical with the derivation for the Euler–Bernoulli beam according to Sect. 2.3.3:

$$\frac{\mathrm{d}Q_z(x)}{\mathrm{d}x} = -q_z(x) , \tag{4.63}$$

$$\frac{\mathrm{d}M_y(x)}{\mathrm{d}x} = +Q_z(x) - m_y(x) . \tag{4.64}$$

Before looking in more detail at the differential equations of the bending line, let us summarize the basic equations for the Levinson beam in Table 4.4. It should be noted that the shear stress and shear strain are now dependent on the x- and z-coordinate.

4.3.4 Differential Equation

Using the expressions for the internal bending moment (4.60) and the internal shear force (4.62) in the equilibrium conditions (4.63) and (4.64) gives finally the coupled systems of partial differential equations for the Levinson beam as:

$$\frac{2}{3}\frac{\partial}{\partial x}\left[GA\left(\phi_y(x) + \frac{\partial u_z(x)}{\partial x}\right)\right] = -q_z(x) , \tag{4.65}$$

$$\frac{1}{5}\frac{\partial}{\partial x}\left[EI_y\left(4\frac{\partial\phi_y(x)}{\partial x}-\frac{\partial^2 u_z(x)}{\partial x^2}\right)\right]-\frac{2}{3}GA\left(\phi_y(x)+\frac{\partial u_z(x)}{\partial x}\right)=-m_y(x).$$

(4.66)

This systems contains—as in the case of the Timoshenko beam—two unknown functions, i.e. the deflection $u_z(x)$ and the cross-sectional rotation $\phi_y(x)$ at $z=0$.

Different formulations of these coupled differential equations are collected in Table 4.5 where different types of loadings, geometry and bedding are differentiated. The last case in Table 4.5 refers again to the elastic or Winkler foundation of a beam, [12]. The elastic foundation modulus k has in the case of beams the unit of force per unit area.

A single-equation description for the Levinson beam can be obtained under the assumption of constant material (E, G) and geometrical (I_y, A) properties: Rearranging and two-times differentiation of Eq. (4.65) gives:

$$\frac{d\phi_y(x)}{dx}=-\frac{d^2 u_z(x)}{dx^2}-\frac{3q_z(x)}{2GA},$$

(4.67)

$$\frac{d^3\phi_y(x)}{dx^3}=-\frac{d^4 u_z(x)}{dx^4}-\frac{3d^2 q_z(x)}{2GAdx^2}.$$

(4.68)

One-time differentiation of Eq. (4.66) gives:

$$\frac{1}{5}EI_y\left(4\frac{d^3\phi_y(x)}{dx^3}-\frac{d^4 u_z(x)}{dx^4}\right)-\frac{2}{3}AG\left(\frac{d^2 u_z(x)}{dx^2}+\frac{d\phi_y(x)}{dx}\right)=-\frac{dm_y(x)}{dx}.$$

(4.69)

Inserting Eq. (4.68) into (4.69) and consideration of (4.67) gives finally the following expression:

$$EI_y\frac{d^4 u_z(x)}{dx^4}=q_z(x)+\frac{dm_y(x)}{dx}-\frac{EI_y}{\frac{5}{6}AG}\frac{d^2 q_z(x)}{dx^2}.$$

(4.70)

The last equation reduces for shear-rigid beams, i.e. $\frac{5}{6}AG \to \infty$, to the classical Euler–Bernoulli formulation as given in Table 2.8.

If we replace in the previous formulations the first- and second-order derivatives, i.e. $\frac{d...}{dx}$ and $\frac{d^2...}{dx^2}$, by formal operator symbols, i.e. the $\mathcal{L}_1$ and $\mathcal{L}_2$–matrices, then the basic equations of the Levinson beam can be stated in a more formal way as given in Table 4.6. Such a matrix formulation is more suitable for the derivation of the principal finite element equation based on the weighted residual method [7, 8].

Under the assumption of constant material (E, G) and geometric (I_y, A) properties, the system of differential equations in Table 4.6 can be solved by using a computer algebra system for constant distributed loads $(q_z(x)=q_0=$ const. and $m_y(x)=m_0=$ const.) to obtain the general analytical solution of the problem, see Fig. 4.7.

Table 4.5 Different formulations of the partial differential equation for a Levinson beam in the x-z plane (x-axis: right facing; z-axis: upward facing)

Configuration	Partial differential equation
E, I_y, A, G	$\dfrac{E I_y}{5}\left(4\dfrac{\partial^2 \phi_y(x)}{\partial x^2} - \dfrac{\partial^3 u_z(x)}{\partial x^3}\right) -$ $\dfrac{2}{3}GA\left(\phi_y(x) + \dfrac{\partial u_z(x)}{\partial x}\right) = 0$ $\dfrac{2GA}{3}\left(\dfrac{\partial \phi_y(x)}{\partial x} + \dfrac{\partial^2 u_z(x)}{\partial x^2}\right) = 0$
$E(x), I_y(x)$ $A(x), G(x)$	$\dfrac{1}{5}\dfrac{\partial}{\partial x}\left[E I_y\left(4\dfrac{\partial \phi_y(x)}{\partial x} - \dfrac{\partial^2 u_z(x)}{\partial x^2}\right)\right] -$ $\dfrac{2}{3}GA\left(\phi_y(x) + \dfrac{\partial u_z(x)}{\partial x}\right) = 0$ $\dfrac{2}{3}\dfrac{\partial}{\partial x}\left[GA\left(\phi_y(x) + \dfrac{\partial u_z(x)}{\partial x}\right)\right] = 0$
$q_z(x)$	$\dfrac{E I_y}{5}\left(4\dfrac{\partial^2 \phi_y(x)}{\partial x^2} - \dfrac{\partial^3 u_z(x)}{\partial x^3}\right) -$ $\dfrac{2}{3}GA\left(\phi_y(x) + \dfrac{\partial u_z(x)}{\partial x}\right) = 0$ $\dfrac{2GA}{3}\left(\dfrac{\partial \phi_y(x)}{\partial x} + \dfrac{\partial^2 u_z(x)}{\partial x^2}\right) = -q_z(x)$
$m_y(x)$	$\dfrac{E I_y}{5}\left(4\dfrac{\partial^2 \phi_y(x)}{\partial x^2} - \dfrac{\partial^3 u_z(x)}{\partial x^3}\right) -$ $\dfrac{2}{3}GA\left(\phi_y(x) + \dfrac{\partial u_z(x)}{\partial x}\right) = -m_y(x)$ $\dfrac{2GA}{3}\left(\dfrac{\partial \phi_y(x)}{\partial x} + \dfrac{\partial^2 u_z(x)}{\partial x^2}\right) = 0$
$k(x)$	$\dfrac{E I_y}{5}\left(4\dfrac{\partial^2 \phi_y(x)}{\partial x^2} - \dfrac{\partial^3 u_z(x)}{\partial x^3}\right) -$ $\dfrac{2}{3}GA\left(\phi_y(x) + \dfrac{\partial u_z(x)}{\partial x}\right) = 0$ $\dfrac{2GA}{3}\left(\dfrac{\partial \phi_y(x)}{\partial x} + \dfrac{\partial^2 u_z(x)}{\partial x^2}\right) = k(x)u_z(x)$

Solution of the coupled DEs for a Levinson beam with AG, EI, q, m = const

```
(%i14)  eqn_1: 2/3*AG*'diff(u(x),x,2) = -q-2/3*AG*'diff(phi(x),x)$
        eqn_2: -1/5*EI*'diff(u(x),x,3)-2/3*AG*'diff(u(x),x) = -m-4/5*EI*'diff(phi(x),x,2)
                                                              +2/3*AG*phi(x)$

        sol : desolve([eqn_1, eqn_2], [u(x), phi(x)])$

        print(" ")$
        print("Equations:")$
        print(" ", eqn_1)$
        print(" ")$
        print(" ", eqn_2)$
        print(" ")$
        print(" ")$
        print("Analytical Solution:")$
        print(ratsimp(sol[1]))$
        print(" ")$
        print(ratsimp(sol[2]))$
```

Equations:

$$\frac{2}{3}AG\left(\frac{d^2}{dx^2}u(x)\right) = -\frac{2}{3}AG\left(\frac{d}{dx}\phi(x)\right) - q$$

$$-\frac{1}{5}EI\left(\frac{d^3}{dx^3}u(x)\right) - \frac{2}{3}AG\left(\frac{d}{dx}u(x)\right) = -\frac{4}{5}EI\left(\frac{d^2}{dx^2}\phi(x)\right) + \frac{2}{3}AG\,\phi(x) - m$$

Analytical Solution:

$$u(x) = -\frac{\begin{array}{c}36AG\,EIx^2\left(\frac{d^2}{dx^2}u(x)\big|_{x=0}\right) + (360AG\,EIx - 40AG^2x^3)\left(\frac{d}{dd}u(x)\big|_{x=0}\right)\\ -144AG\,EIx^2\left(\frac{d}{dx}\phi(x)\big|_{x=0}\right) + 15AGqx^4\\ + (60AGm - 40\,\phi(0)AG^2)x^3 - 216EIqx^2 + 360\,u(0)AG\,EI\end{array}}{360AGEI}$$

$$\phi(x) = -\frac{\begin{array}{c}6AG\,EIx\left(\frac{d^2}{dx^2}u(x)\big|_{x=0}\right) - 10AG^2x^2\left(\frac{d}{dx}u(x)\big|_{x=0}\right)\\ -24AG\,EIx\left(\frac{d}{dx}\phi(x)\big|_{x=0}\right) + 5AGqx^3\\ + (15AGm - 10\,\phi(0)AG^2)x^2 + 9EIqx - 30\,\phi(0)AG\,EI\end{array}}{30AGEI}$$

Fig. 4.7 Solution of the coupled system of differential equations for the Levinson beams based on the computer algebra system Maxima

Table 4.6 Different formulations of the basic equations for a Levinson beam (bending in the x-z plane; x-axis along the principal beam axis). E: Young's modulus; G: shear modulus; A: cross-sectional area; I_y: second moment of area; q_z: length-specific distributed force; m_y: length-specific distributed moment; e: generalized strains; s^*: generalized stresses

Specific formulation	General formulation [1]
Kinematics	
$\begin{bmatrix} \frac{du_y}{dx} + \phi_y \\ -\frac{d^2 u_y}{dx^2} + 4\frac{d\phi_y}{dx} \end{bmatrix} = \begin{bmatrix} \frac{d}{dx} & +1 \\ -\frac{d^2}{dx^2} & \frac{d}{dx} \end{bmatrix} \begin{bmatrix} u_z \\ \phi_y \end{bmatrix}$	$e = \mathcal{L}_{2^*} u$
Constitution	
$\begin{bmatrix} Q_z \\ M_y \end{bmatrix} = \begin{bmatrix} \frac{2}{3}AG & 0 \\ 0 & \frac{1}{5}EI_y \end{bmatrix} \begin{bmatrix} \frac{du_z}{dx} + \phi_y \\ -\frac{d^2 u_z}{dx^2} + 4\frac{d\phi_y}{dx} \end{bmatrix}$	$s^* = D^* e$
Equilibrium	
$\begin{bmatrix} \frac{d}{dx} & 0 \\ -1 & \frac{d}{dx} \end{bmatrix} \begin{bmatrix} Q_z \\ M_y \end{bmatrix} + \begin{bmatrix} q_z \\ m_y \end{bmatrix} = \begin{bmatrix} 0 \\ 0 \end{bmatrix}$	$\mathcal{L}_{1^*}^{\mathrm{T}} s^* + b^* = 0$
PDE	
$\dfrac{d}{dx}\left[\frac{2}{3}GA\left(\frac{du_z}{dx} + \phi_y \right) \right] + q_z = 0$ $\dfrac{d}{dx}\left[\frac{1}{5}EI_y\left(-\frac{d^2 u_z}{dx^2} + 4\frac{d\phi_y}{dx} \right) \right] -$ $\quad - \frac{2}{3}GA\left(\frac{du_z}{dx} + \phi_y \right) + m_y = 0$	$\mathcal{L}_{1^*}^{\mathrm{T}} D^* \mathcal{L}_{2^*} u + b^* = 0$

The general analytical solution provided in Fig. 4.7, i.e.,

$$
u_z(x) = -\frac{\begin{aligned} &36AG\,EI_y x^2 \left(\tfrac{d^2 u_z(0)}{dx^2} \right) + \left(360AG\,EI_y x - 40(AG)^2 x^3 \right) \left(\tfrac{du_z(0)}{dx} \right) \\ &- 144AG\,EI_y x^2 \left(\tfrac{d\phi_y(0)}{dx} \right) + 15AG q_0 x^4 \\ &+ \left(60AGm - 40\phi_y(0)(AG)^2 \right) x^3 - 216EI_y q_0 x^2 + 360 u_z(0)AG\,EI_y \end{aligned}}{360 AG EI_y},
$$

$$
\phi_y(x) = \frac{\begin{aligned} &6AG\,EI_y x \left(\tfrac{d^2 u_z(0)}{dx^2} \right) - 10(AG)^2 x^2 \left(\tfrac{du_z(0)}{dx} \right) - 24AG\,EI_y x \left(\tfrac{d\phi_y(0)}{dx} \right) \\ &+ 5AG q_0 x^3 + \left(15AG m_0 - 10\phi_y(0)(AG)^2 \right) x^2 \\ &+ 9EI_y q_0 x - 30\phi_y(0)AG\,EI_y \end{aligned}}{30 AG EI_y},
$$

where $\frac{du_z(0)}{dx}$, $\frac{d\phi_y(0)}{dx}$, $u_z(0)$, and $\phi_y(0)$ are the four boundary values to adjust the general solution to a particular problem, can be represented in a slightly different way based on a different set of constants:

$$u_z(x) = \frac{1}{EI_y}\left(\frac{q_0x^4}{24} + \frac{x^3}{6}\underbrace{\left[-\frac{2}{3}AG\frac{du_z(0)}{dx} - \frac{2}{3}AG\phi_y(0)\right]}_{c_1} + \right.$$

$$+ \frac{x^2}{2}\underbrace{\left[\frac{EI_y\,d^2u_z(0)}{5\;dx^2} - \frac{4EI_y\,d\phi_y(0)}{5\;dx} - \frac{6EI_yq_0}{5\;AG}\right]}_{c_2} + \underbrace{EI_y\frac{du_z(0)}{dx}}_{c_3}x +$$

$$\left.+ \underbrace{EI_yu_z(0)}_{c_4}\right) + \frac{m_0x^3}{6EI_y}, \tag{4.71}$$

$$\phi_y(x) = -\frac{1}{EI_y}\left(\frac{q_0x^3}{3} + \frac{x^2}{2}\underbrace{\left[-\frac{2}{3}AG\frac{du_z(0)}{dx} - \frac{2}{3}AG\phi_y(0)\right]}_{c_1} + \right.$$

$$\left.+ x\underbrace{\left[\frac{EI_y\,d^2u_z(0)}{5\;dx^2} - \frac{4EI_y\,d\phi_y(0)}{5\;dx} - \frac{6EI_yq_0}{5\;AG}\right]}_{c_2} + \underbrace{EI_y\frac{du_z(0)}{dx}}_{c_3}\right)$$

$$- \frac{q_0x}{\frac{2}{3}AG} + \underbrace{\left[\frac{du_z(0)}{dx} + \phi_y(0)\right]}_{\frac{-c_1}{\frac{2}{3}AG}} - \frac{m_0x^2}{2EI_y}. \tag{4.72}$$

The general expressions for the internal bending moment and shear force distributions can be obtained under consideration of Eqs. (4.60) and (4.62). Thus, we can finally state the general solution for constant material and geometrical properties as well as constant distributed loads as:

$$u_z(x) = \frac{1}{EI_y}\left(\frac{q_0 x^4}{24} + c_1\frac{x^3}{6} + c_2\frac{x^2}{2} + c_3 x + c_4\right) + \frac{m_0 x^3}{6EI_y}, \tag{4.73}$$

$$\phi_y(x) = -\frac{1}{EI_y}\left(\frac{q_0 x^3}{6} + c_1\frac{x^2}{2} + c_2 x + c_3\right) - \frac{q_0 x}{\frac{2}{3}AG} - \frac{c_1}{\frac{2}{3}AG} - \frac{m_0 x^2}{2EI_y}, \tag{4.74}$$

$$M_y(x) = -\left(\frac{q_0 x^2}{2} + c_1 x + c_2\right) - \frac{\frac{4}{5}q_0 EI_y}{\frac{2}{3}AG} - m_0 x, \tag{4.75}$$

$$Q_z(x) = -(q_0 x + c_1), \tag{4.76}$$

where the four constants of integration $c_i\,(i = 1, \ldots, 4)$ must be determined based on the boundary conditions.

4.4 Comparison of Both Planes

The comparison of the elementary basic equations for bending in the x-y and x-y plane are shown in Table 4.7. All the basic equations have the same structure but the different definition of a positive rotation yields to some different signs.

4.5 Unsymmetrical Bending in Both Planes

The line of reasoning presented in Sects. 2.4 and (3.5) is again followed in the following. The total normal strain in the x-direction is obtained by simple superposition of its components from the bending contributions in the x-y and x-z plane (see Eqs. (4.25) and (4.51) for details), i.e.

$$\varepsilon_x(x, y, z) = \varepsilon_x^{x-y}(x, y) + \varepsilon_x^{x-z}(x, z). \tag{4.77}$$

Thus, Hooke's law according to Eqs. (4.30) or (4.30) can be generalized to the following formulation (see Eqs. (4.25) and (4.51)):

$$\sigma_x(x, y, z) = E\varepsilon_x(x, y, z) \tag{4.78}$$

$$= E\left(-y\frac{d\phi_z(x)}{dx} + y^3\frac{d\theta^{x-y}(x)}{dx} + z\frac{d\phi_y(x)}{dx} + z^3\frac{d\theta^{x-z}(x)}{dx}\right). \tag{4.79}$$

Based on this stress distribution, the evaluation of the internal bending moments $M_z(x)$ and $M_y(x)$ (see Eqs. (4.32) and (4.58)) has to be performed. In the case, for example, of the bending moment $M_z(x)$, the following integral must be considered:

Table 4.7 Elementary basic equations for the bending of a Levinson beam in the x-z and x-y plane. The differential equations are given under the assumption of constant material (E, G) and geometrical (I, A) properties

Equation	x-z plane	x-y plane
Kinematics	$\varepsilon_x(x,z) = z\frac{d\phi_y(x)}{dx} + z^3\frac{d\theta(x)}{dx}$	$\varepsilon_x(x,y) = -y\frac{d\phi_z(x)}{dx} + y^3\frac{d\theta(x)}{dx}$
	$\gamma_{xz}(x,z) =$ $\frac{h^2-4z^2}{h^2}\left(\phi_y(x) + \frac{\partial u_z(x)}{\partial x}\right)$	$\gamma_{xy}(x,y) =$ $\frac{h^2-4y^2}{h^2}\left(-\phi_z(x) + \frac{\partial u_y(x)}{\partial x}\right)$
Equilibrium	$\frac{dQ_z(x)}{dx} = -q_z(x)$	$\frac{dQ_y(x)}{dx} = -q_y(x)$
	$\frac{dM_y(x)}{dx} = +Q_z(x) - m_y(x)$	$\frac{dM_z(x)}{dx} = -Q_y(x) - m_z(x)$
Constitution	$\sigma_x(x,z) = E\varepsilon_x(x,z)$	$\sigma_x(x,y) = E\varepsilon_x(x,y)$
	$\tau_{xz}(x,z) = G\gamma_{xz}(x,z)$	$\tau_{xy}(x,z) = G\gamma_{xy}(x,y)$
	$M_y(x) =$ $\frac{EI_y}{5}\left(4\frac{\partial\phi_y(x)}{\partial x} - \frac{\partial^2 u_z(x)}{\partial x^2}\right)$	$M_z(x) =$ $\frac{EI_z}{5}\left(4\frac{\partial\phi_z(x)}{\partial x} + \frac{\partial^2 u_y(x)}{\partial x^2}\right)$
	$Q_z(x) = \frac{2}{3}GA\left(\phi_y(x) + \frac{\partial u_z(x)}{\partial x}\right)$	$Q_y(x) =$ $\frac{2}{3}GA\left(-\phi_z(x) + \frac{\partial u_y(x)}{\partial x}\right)$
PDEs	$\frac{EI_y}{5}\left(4\frac{\partial^2\phi_y(x)}{\partial x^2} - \frac{\partial^3 u_z(x)}{\partial x^3}\right)$ $-\frac{2}{3}GA\left(\phi_y(x) + \frac{\partial u_z(x)}{\partial x}\right) =$ $-m_y(x)$ $\frac{2GA}{3}\left(\frac{\partial\phi_y(x)}{\partial x} + \frac{\partial^2 u_z(x)}{\partial x^2}\right) =$ $-q_z(x)$	$\frac{EI_z}{5}\left(4\frac{\partial^2\phi_z(x)}{\partial x^2} + \frac{\partial^3 u_y(x)}{\partial x^3}\right)$ $+\frac{2}{3}GA\left(-\phi_z(x) + \frac{\partial u_y(x)}{\partial x}\right) =$ $-m_z(x)$ $\frac{2GA}{3}\left(-\frac{\partial\phi_z(x)}{\partial x} + \frac{\partial^2 u_y(x)}{\partial x^2}\right) =$ $-q_y(x)$

$$M_z(x) = \int_A -y\sigma_x dA \tag{4.80}$$

$$= -E\int_A \left(-y^2\frac{d\phi_z(x)}{dx} + y^4\frac{d\theta^{x-y}(x)}{dx} + yz\frac{d\phi_y(x)}{dx} + yz^3\frac{d\theta^{x-z}(x)}{dx}\right) dA . \tag{4.81}$$

Since major parts of the derivations in the previous sections (see for example Eq. (4.54)), are based on the assumption of a rectangular cross section ($b \times h$), the integration over the third and fourth term in Eq. (4.81), i.e. 'yz' and 'yz^3', results in zero and thus a decoupling of the bending deformation in the x-y and x-z plane for a symmetrical section.

4.6   Supplementary Problems

4.1 Cantilever Levinson beam with different end loads and deformations

Calculate the analytical solution for the deflection $u_z(x)$ and rotation $\phi_y(x)$ of the cantilever Levinson beams shown in Fig. 4.8. It can be assumed for this exercise that the bending $(E I_y)$ and the shear (AG) stiffnesses are constant. Calculate in addition for all four cases the reactions at the fixed support and the distributions of the bending moment and shear force.

4.2 Cantilever Levinson beam with point load: effective stress over cross section

Given is the cantilever Levinson beam with an end load F_0 as shown in Fig. 4.9. Calculate the effective stress based on the normal and shear stress distributions (see Fig. 4.1c) as a function of the vertical z-coordinate.

Based on the von Mises effective stress [6], i.e.,

$$\sigma_{\text{eff}}(\sigma, \tau) = \sqrt{\sigma^2 + 3\tau^2}, \qquad (4.82)$$

determine the location of the maximum effective stress.

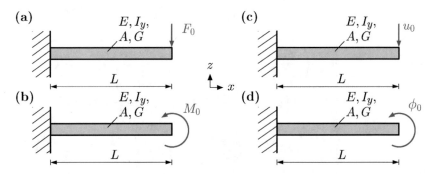

Fig. 4.8 Cantilever Levinson beam with different end loads and deformations: **a** single force; **b** single moment; **c** displacement; **d** rotation

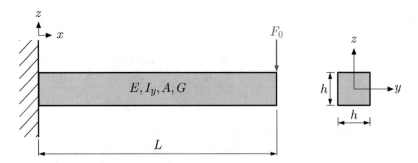

Fig. 4.9 Cantilever Levinson beam with point load

References

1. Altenbach H, Öchsner A (eds) (2020) Encyclopedia of continuum mechanics. Springer, Berlin
2. Elishakoff I, Irretier H (eds) (1987) Refined dynamical theories of beams, plates and shells and their applications: proceedings of the euromech-colloquium 219. Springer, Berlin
3. Ghugal YM, Shimpi RP (2001) A review of refined shear deformation theories for isotropic and anisotropic laminated beams. J Reinf Plast Comp 20:255–272
4. Levinson M (1981) A new rectangular beam theory. J Sound Vib 74:81–87
5. Öchsner A (2014) Elasto-plasticity of frame structure elements: modeling and simulation of rods and beams. Springer, Berlin
6. Öchsner A (2016) Continuum damage and fracture mechanics. Springer, Singapore
7. Öchsner A, Merkel M (2018) One-dimensional finite elements: an introduction to the FE method. Springer, Cham
8. Öchsner A (2020) Computational statics and dynamics: an introduction based on the finite element method. Springer, Singapore
9. Reddy JN (1990) A general non-linear third-order theory of plates with moderate thickness. Int J Nonlinear Mech 25:677–686
10. Sayyad AS (2011) Comparison of various refined beam theories for the bending and free vibration analysis of thick beams. Appl Comput Mech 5:217–230
11. Wang CM, Reddy JN, Lee KH (2000) Shear deformable beams and plates: relationships with classical solution. Elsevier, Oxford
12. Winkler E (1867) Die Lehre von der Elasticität und Festigkeit mit besonderer Rücksicht auf ihre Anwendung in der Technik. H. Dominicus, Prag

Chapter 5
Comparison of the Approaches

Abstract This chapter investigates the influence of the different beam theories, i.e., Euler–Bernoulli, Timoshenko, and Levinson, on the deformations of beams. Some typical cases in regards to support and loading conditions are considered and the maximum beam deformation is presented as a function of the slenderness ratio.

5.1 Introduction

Chapters 2, 3 and 4 introduced the beam theories according to Euler–Bernoulli, Timoshenko and Levinson. The following derivations are investigating the influence of these theories on the maximum deflection of beams. In particular, the difference between the theories for different levels of 'thinness' or 'thickness' of the cross section in relation to the beam length is evaluated. Let us look, as an example, on the three different beam configurations as shown in Fig. 5.1, i.e. a cantilever configuration with point load, a cantilever configuration with distributed load, and a simply supported beam with a point load, see [1–3] for further details. It is assumed for simplicity that the material (E, G) and geometrical properties (I_y, k_s, A) are constant.

The boundary conditions for the case shown in Fig. 5.1a can be stated as

$$u_z(0) = 0 \,, \qquad\qquad M_y(L) = 0 \,, \qquad\qquad (5.1)$$

$$\phi_y(0) = 0 \,, \qquad\qquad Q_z(L) = F_0 \,, \qquad\qquad (5.2)$$

which allow to determine the constants of integration in the general solution for the Euler–Bernoulli beam (EB) according to Eqs. (2.77)–(2.80) as $c_1 = -F_0$, $c_2 = F_0 L$, $c_3 = 0$, and $c_4 = 0$. Thus, the bending line can be expressed as

$$u_z^{EB}(x) = \frac{1}{EI_y}\left(-\frac{F_0 x^3}{6} + \frac{F_0 L x^2}{2}\right) . \qquad\qquad (5.3)$$

A. Öchsner, *Classical Beam Theories of Structural Mechanics*,
https://doi.org/10.1007/978-3-030-76035-9_5

Fig. 5.1 Beams under different support and loading conditions: **a** cantilever configuration with point load, **b** cantilever configuration with distributed load, and **c** simply supported with point load

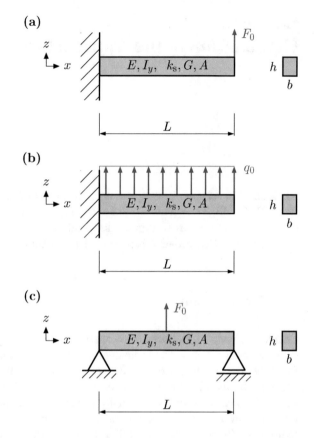

The maximum deflection is obtained for $x = L$ and Eq. (5.3) can be simplified to:

$$u_z^{\text{EB}}(x) = \frac{F_0 L^3}{3EI_y}.$$ (5.4)

The in Eqs. (5.1)–(5.2) stated boundary conditions do not depend on the applied beam theory. Thus, they can be used the same way for the Timoshenko beam formulation and allow the determination of the constants of integration in the general solution for the Timoshenko (T) beam according to Eqs. (3.93)–(3.96) as $c_1 = -F_0$, $c_2 = F_0 L$, $c_3 = \frac{EI_y F_0}{k_s AG}$, and $c_4 = 0$. Thus, the maximum deflection at $x = L$ can be expressed as

$$u_z^{\text{T}}(L) = \frac{F_0 L^3}{3EI_y} + \frac{F_0 L^3}{3EI_y} \times \frac{3EI_y}{k_s AGL^2},$$ (5.5)

or under consideration of $k_s = 5/6$ and $A = 12I_y/h^2$ in normalized representation:

Table 5.1 Constants of integration and normalized deflection for the case shown in Fig. 5.1a, i.e., a cantilever beam with a point load F_0 at $x = L$

Theory	c_1	c_2	c_3	c_4	$\dfrac{u_z(L)}{\frac{F_0 L^3}{3EI_y}}$
Euler–Bernoulli	$-F_0$	$F_0 L$	0	0	1
Timoshenko	$-F_0$	$F_0 L$	$\dfrac{EI_y F_0}{k_s AG}$	0	$1 + \dfrac{3(1+v)}{5}\left(\dfrac{h}{L}\right)^2$
Levinson	$-F_0$	$F_0 L$	$\dfrac{3EI_y F_0}{2AG}$	0	$1 + \dfrac{3(1+v)}{4}\left(\dfrac{h}{L}\right)^2$

$$\frac{u_z^{\mathrm{T}}(L)}{\frac{F_0 L^3}{3EI_y}} = 1 + \frac{3(1+v)}{5}\left(\frac{h}{L}\right)^2 . \tag{5.6}$$

Since there is no distributed load applied ($q_0 = 0$) for the case in Fig. 5.1a, we can obtain the solution for the Levinson beam from the Timoshenko solution by setting $k_s \to 2/3$, see Eqs. (4.73)–(4.76):

$$\frac{u_z^{\mathrm{L}}(L)}{\frac{F_0 L^3}{3EI_y}} = 1 + \frac{3(1+v)}{4}\left(\frac{h}{L}\right)^2 . \tag{5.7}$$

The major results for the cantilever beam with a point load at $x = L$ are summarized in Table 5.1.

The boundary conditions for the second case shown in Fig. 5.1b can be stated as

$$u_z(0) = 0 , \qquad\qquad M_y(L) = 0 , \tag{5.8}$$
$$\phi_y(0) = 0 , \qquad\qquad Q_z(L) = 0 . \tag{5.9}$$

This allows us to determine the constants of integration in the general solution for the Euler–Bernoulli beam (EB) according to Eqs. (2.77)–(2.80) as $c_1 = -q_0 L$, $c_2 = q_0 L^2/2$, $c_3 = 0$, and $c_4 = 0$ and the maximum deflection at $x = L$ can be expressed as

$$u_z^{\mathrm{EB}}(L) = \frac{q_0 L^4}{8EI_y} . \tag{5.10}$$

For the same set of boundary conditions, the constants of integration in the general solution for the Timoshenko (T) beam according to Eqs. (3.93)–(3.96) can be derived as $c_1 = -q_0 L$, $c_2 = q_0 L^2/2 - q_0 EI_y/(k_s AG)$, $c_3 = EI_y q_0 L/(k_s AG)$, and $c_4 = 0$.

Thus, the maximum deflection at $x = L$ can be expressed as

$$u_z^{\mathrm{T}}(L) = \frac{q_0 L^4}{8EI_y} + \frac{q_0 L^4}{8EI_y} \times \frac{8EI_y}{k_s AGL^2}, \tag{5.11}$$

or under consideration of $k_s = 5/6$ and $A = 12I_y/h^2$ in normalized representation:

$$\frac{u_z^{\mathrm{T}}(L)}{\frac{q_0 L^4}{8EI_y}} = 1 + \frac{4(1+v)}{5}\left(\frac{h}{L}\right)^2. \tag{5.12}$$

The determination of the constants of integration for the Levinson beam must consider in this case ($q_0 \neq 0$) that a simple transformation ($k_s \to 2/3$) of the Timoshenko constants is not advisable since Eqs. (3.95) and (4.75) do not only differ by k_s and $2/3$ but also by a factor of $4/5$ in the fraction with q_0. The correct evaluation of the constants of integration for the Levinson beams yields: $c_1 = -q_0 L$, $c_2 = q_0 L^2/2 - 6q_0 EI_y/(5AG)$, $c_3 = EI_y q_0 L/(k_s AG)$, and $c_4 = 0$. Thus, the maximum deflection at $x = L$ can be expressed in normalized form as

$$\frac{u_z^{\mathrm{L}}(L)}{\frac{q_0 L^4}{8EI_y}} = 1 + \frac{6(1+v)}{5}\left(\frac{h}{L}\right)^2. \tag{5.13}$$

The major results for the cantilever beam with a constant distributed load are summarized in Table 5.2.

The boundary conditions for the third case shown in Fig. 5.1c can be stated as

Table 5.2 Constants of integration and normalized deflection for the case shown in Fig. 5.1b, i.e., a cantilever beam with a constant distributed load q_0

Theory	c_1	c_2	c_3	c_4	$\dfrac{u_z(L)}{\frac{q_0 L^4}{8EI_y}}$
Euler–Bernoulli	$-q_0 L$	$\dfrac{q_0 L^2}{2}$	0	0	1
Timoshenko	$-q_0 L$	$\dfrac{q_0 L^2}{2} - \dfrac{EI_y q_0}{k_s AG}$	$\dfrac{EI_y q_0 L}{k_s AG}$	0	$1 + \dfrac{4(1+v)}{5}\left(\dfrac{h}{L}\right)^2$
Levinson	$-q_0 L$	$\dfrac{q_0 L^2}{2} - \dfrac{6EI_y q_0}{5AG}$	$\dfrac{EI_y q_0 L}{\frac{2}{3}AG}$	0	$1 + \dfrac{6(1+v)}{5}\left(\dfrac{h}{L}\right)^2$

$$u_z(0) = 0, \qquad\qquad M_y(0) = 0, \qquad\qquad (5.14)$$

$$\phi_y(L/2) = 0, \qquad\qquad Q_z(0) = F_0/2. \qquad\qquad (5.15)$$

This allows us to determine the constants of integration in the general solution for the Euler–Bernoulli beam (EB) according to Eqs. (2.77)–(2.80) as $c_1 = -F_0/2$, $c_2 = 0$, $c_3 = F_0 L^2/16$, and $c_4 = 0$ and the maximum deflection at $x = L/2$ can be expressed as

$$u_z^{EB}(L/2) = \frac{F_0 L^3}{48 E I_y}. \qquad\qquad (5.16)$$

For the same set of boundary conditions, the constants of integration in the general solution for the Timoshenko (T) beam according to Eqs. (3.93)–(3.96) can be derived as $c_1 = -F_0/2$, $c_2 = 0$, $c_3 = E I_y F_0/(2k_s AG) + F_0 L^2/16$, and $c_4 = 0$. Thus, the maximum deflection at $x = L/2$ can be expressed as

$$u_z^T(L/2) = \frac{F_0 L^3}{48 E I_y} + \frac{F_0 L^3}{48 E I_y} \times \frac{48 E I_y}{4 k_s AG L^2}, \qquad\qquad (5.17)$$

or under consideration of $k_s = 5/6$ and $A = 12 I_y/h^2$ in normalized representation:

$$\frac{u_z^T(L/2)}{\frac{F_0 L^3}{48 E I_y}} = 1 + \frac{12(1+v)}{5}\left(\frac{h}{L}\right)^2. \qquad\qquad (5.18)$$

Since there is no distributed load applied ($q_0 = 0$) for the case in Fig. 5.1c, we can obtain the solution for the Levinson beam from the Timoshenko solution by setting $k_s \to 2/3$, see Eqs. (4.73)–(4.76):

$$\frac{u_z^L(L/2)}{\frac{F_0 L^3}{48 E I_y}} = 1 + \frac{3(1+v)}{1}\left(\frac{h}{L}\right)^2. \qquad\qquad (5.19)$$

The major results for the simply supported beam with a point load at $x = L/2$ are summarized in Table 5.3.

The normalized maximum deflections according to Tables 5.1, 5.2 and 5.3 are compared in Fig. 5.2. It becomes obvious that the difference between the Bernoulli, i.e. the thin beam formulation, and the Timoshenko/Levinson beams, i.e., the thick beam formulations, becomes smaller and smaller for a decreasing slenderness ratio, meaning for beams at which the length L is significantly larger compared to the height h. In addition, it can be seen that the Levinson formulation gives for all cases a slighly larger deformation than the Timoshenko approach. This difference is increasing for increasing slenderness ratio. Finally, these diagrams allow to evaluate the classical rule of thumb that a beam is considerd as 'thin' if the length L is approximately ten times a characteristic dimension of the cross section (for example the height h).

Fig. 5.2 Maximum beam deflections according to the theories of Euler–Bernoulli, Timoshenko and Levinson: **a** cantilever configuration with point load, **b** cantilever configuration with distributed load, and **c** simply supported with point load

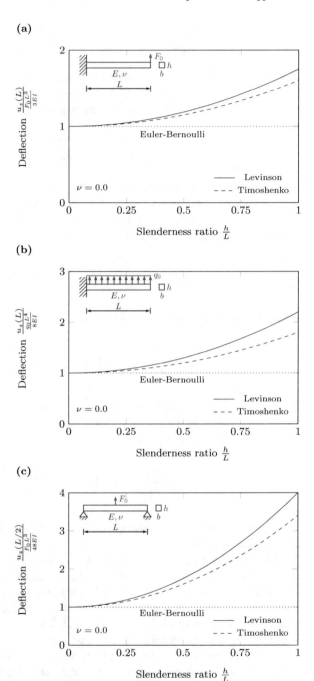

5.2 Supplementary Problems

5.1 Maximum beam deflections according to the Levinson and Timoshenko theory for different Poisson's ratios

Reconsider the beam problems shown in Figs. 5.1 and 5.2 and evaluate the influence of Poisson's ratio on the maximum deformation.

5.2 Maximum beam deflections according to the Timoshenko theory for beams with circular cross sections

The cantilever Timoshenko beam shown in Fig. 5.3 is either loaded by a single force F_0 at its right-hand end or by a constant distributed load q_0. The bending stiffness EI_y and the shear stiffness $k_s AG$ are constant, the total length of the beam is equal to

Table 5.3 Constants of integration and normalized deflection for the case shown in Fig. 5.1c, i.e., a simply supported beam with a point load F_0 at $x = L/2$

Theory	c_1	c_2	c_3	c_4	$\dfrac{u_z(L/2)}{\frac{F_0 L^3}{48 E I_y}}$
Euler–Bernoulli	$-\dfrac{F_0}{2}$	0	$\dfrac{F_0 L^2}{16}$	0	1
Timoshenko	$-\dfrac{F_0}{2}$	0	$\dfrac{EI_y F_0}{2k_s AG} + \dfrac{F_0 L^2}{16}$	0	$1 + \dfrac{12(1+v)}{5}\left(\dfrac{h}{L}\right)^2$
Levinson	$-\dfrac{F_0}{2}$	0	$\dfrac{EI_y F_0}{\frac{4}{3} AG} + \dfrac{F_0 L^2}{16}$	0	$1 + \dfrac{3(1+v)}{1}\left(\dfrac{h}{L}\right)^2$

Fig. 5.3 Cantilever Timoshenko beam: **a** single force case and **b** distributed load case

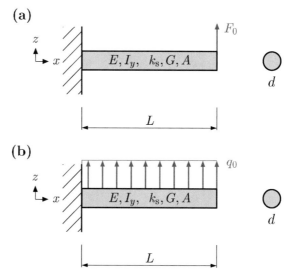

L, and the circular cross section has a diameter of d. Assume $k_s = \frac{5}{6}$ for the circular cross section. Determine the expressions of the bending lines $(u_z(x))$ and sketch the deflections of the right-hand end $(x = L)$ as a function of the slenderness ratio $\frac{d}{L}$ for $\nu = 0.0, 0.3$, and 0.5. Compare the results with the classical Euler–Bernoulli solutions for thin beams.

References

1. Öchsner A (2014) Elasto-plasticity of frame structure elements: modelling and simulation of rods and beams. Springer, Berlin
2. Öchsner A (2020) Computational statics and dynamics: an introduction based on the finite element method. Springer, Singapore
3. Öchsner A (2020) Stoff- und Formleichtbau: Leichter Einstieg mit eindimensionalen Strukturen. Springer, Wiesbaden

Chapter 6
Outlook: Finite Element Approach

Abstract This chapter briefly introduces the finite element method, i.e. a mathematical procedure to approximately solve the differential equations which were presented in the previous chapters. The application is discussed in the context of bending members and the advantage of solving different field problems based on the same discretization is highlighted.

The simple analytical solutions for the different beam theories according to Eqs. (2.37)–(2.40), (3.45)–(3.48), and (4.47)–(4.50) can only be applied to simple problems. Nowadays, numerical approximation methods have established themselves for more complex structures and questions, whereby the finite element method (FEM) represents the standard tool in the field of structural mechanics. The basic idea of this approximation is that the basic equations of continuum mechanics (see Fig. 1.1) are no longer fulfilled for every point on the continuum, but only averaged for a so-called finite element. Deformations are only calculated at a finite number of points, the so-called nodes, and simply interpolated between them. These nodes are placed at least at the ends/corners of the elements and allow individual elements to be connected to form a coherent structure, the so-called finite element mesh. Thus, a structural part is approximated by a mesh of connected elements and the degrees of freedom of the system are reduced to the nodes. More information and details on the finite element method can be found in the relevant literature, see for example [2, 4, 5].

A beam can, for example, be approximated by a one-dimensional beam element as part of a finite element analysis, see Fig. 6.1.

The beam element itself is described here by two nodes and a connecting line. Cross-sections and material properties, in the case of one-dimensional elements, are only provided as numerical input in a finite element program and do not have to be taken into account in the mesh. This results in an extremely simplified generation of a beam structure in a finite element program. The following is a brief example of an axial extensometer, see Fig. 6.2a. As part of the so-called modeling, the real structure from Fig. 6.2a must first be converted into a mechanical model through simplifications and assumptions, see Fig. 6.2b. Then the simplified structure is converted into a finite element mesh composed of one-dimensional beam elements (so-called

A. Öchsner, *Classical Beam Theories of Structural Mechanics*,
https://doi.org/10.1007/978-3-030-76035-9_6

Fig. 6.1 a Beam structure with I-section; **b** approximation using a single beam element. The element nodes are symbolized by circles (∘)

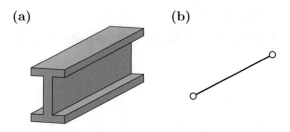

Fig. 6.2 a Flat tensile specimen with axial extensometer; **b** mechanical model of the axial extensometer under consideration of symmetry, **c** finite element mesh consisting of two elements I and II

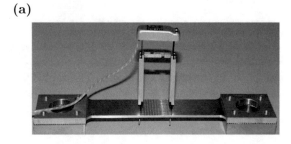

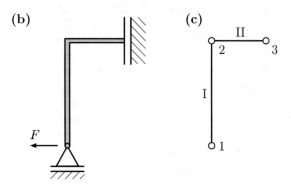

discretization). The beam elements in commercial programs are so-called generalized beams, which can also deform in the axial direction or can be twisted along the longitudinal axis.

For each of these beam elements, the so-called principal finite element equation $K^e u^e = f^e$ can be given at the element level. The so-called stiffness matrix K^e contains the information about the material and the geometry of the element. The column matrix of the unknowns u^e includes the deformations, i.e. the displacements and rotations at the nodes, and the load column matrix f^e contains the external forces and moments acting on the element. In general, the accuracy of a finite element analysis increases with the number of elements. The main equations for the individual elements can be assembled to a global system of equations in the following form:

$$Ku = f .$$ (6.1)

After considering the boundary conditions, the solution for this linear system of equations is:

$$u = K^{-1}f.$$ (6.2)

The finite element method still allows other questions to be answered based on the same mesh. Taking into account the structural masses, which are summarized in a mass matrix M, the principal finite element equation results as follows:

$$M\ddot{u}(t) + Ku(t) = f(t).$$ (6.3)

The solution of this equation in the time domain is often done using classical difference methods and provides the deformation $u(t)$, the velocity $\dot{u}(t)$ and the acceleration $\ddot{u}(t)$ of the transient problem.

The consideration of the same finite element model allows the determination of the eigenfrequencies and eigenmodes by solving an eigenvalue problem:

$$\det\left(K - \omega_i M\right) = 0,$$ (6.4)

whereas the ω_i represent the eigenfrequencies of the system. The eigenmodes, i.e. the deformed shape of the structure at certain natural frequencies, are defined by the following relationship:

$$\left(K - \omega_i^2 M\right)\Phi = 0,$$ (6.5)

whereas Φ represents the eigenmodes of the system. Furthermore, the solution of a further eigenvalue problem (with the same mesh) allows the determination of the buckling load within the framework of a stability analysis:

$$\det\left(K\right) = \det\left(K^{\mathrm{el}} + \lambda K^{\mathrm{geo}}\right) = 0,$$ (6.6)

where K^{el} is the stiffness matrix according to Eq. (6.2) for a linear-elastic problem and K^{geo} represents the geometric stiffness matrix including the external load, see [2]. The buckling load finally results from the relation λf.

The above-described solutions to various structural-mechanical problems can be performed in a commercial finite element program [3] based on the same mesh, simply by switching to the different solution modes. For more particular problems, there are commercial finite element packages which allow to expand the functionality based on user subroutines, see [1] for details.

References

1. Javanbakht Z, Öchsner A (2017) Advanced finite element simulation with MSC Marc: application of user subroutines. Springer, Cham
2. Öchsner A, Merkel M (2018) One-dimensional finite elements: an introduction to the FE method. Springer, Cham
3. Öchsner A, Öchsner M (2018) A first introduction to the finite element analysis program MSC Marc/Mentat. Springer, Cham
4. Öchsner A (2020) Computational statics and dynamics: an introduction based on the finite element method. Springer, Singapore
5. Reddy JN (2006) An introduction to the finite element method. McGraw Hill, Singapore

Chapter 7
Answers to Supplementary Problems

Abstract This chapter provides the solutions to the supplementary problems given at the end of chapters 2 through 5. Where appropriate, intermediate steps are provided to better understand the solution procedure.

7.1 Problems from Chap. 2

2.1 Cantilever beam with different end loads and deformations

Case (a): Single force F_0 at $x = L$

Let us start the solution procedure by sketching the free-body diagram as shown in Fig. 7.1a.

The consideration of the global force and moment equilibrium would allow to calculate the reactions at the fixed support, i.e., at $x = 0$:

$$\sum_i F_{z_i} = 0 \quad \Leftrightarrow \quad F_z^R(0) - F_0 = 0 \quad \Rightarrow \quad F_z^R(0) = F_0 , \tag{7.1}$$

$$\sum_i M_{y_i} = 0 \quad \Leftrightarrow \quad M_y^R(0) + F_0 L = 0 \quad \Rightarrow \quad M_y^R(0) = -F_0 L . \tag{7.2}$$

The boundary conditions can be stated at the left-hand end as

$$u_z(0) = 0 \quad \text{and} \quad \varphi_y(0) = 0 , \tag{7.3}$$

while the consideration of the force and moment equilibrium at the right-hand boundary (see Fig. 7.2) requires that

$$Q_z(L) = -F_0 \quad \text{and} \quad M_y(L) = 0 . \tag{7.4}$$

Consideration of the boundary condition $(7.3)_1$ in the general expression for the displacement distribution (2.77) gives the fourth constant of integration as: $c_4 = 0$.

(a)

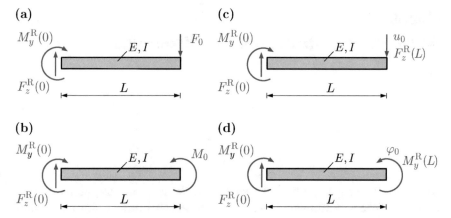

(c)

(b)

(d)

Fig. 7.1 Free-body diagrams of the cantilever beams with different end loads and deformations: **a** single force; **b** single moment; **c** displacement; **d** rotation

Fig. 7.2 Equilibrium between internal reactions and external load at $x = L$

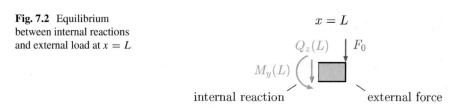

In a similar way, the third constant of integration can be obtained by considering the boundary condition $(7.3)_2$ in the general expression for the rotation distribution (2.80): $c_3 = 0$. Introducing the boundary conditions at the right-hand end, i.e. Eq. (7.4) in the expressions for the bending moment and shear force according to Eqs. (2.78) and (2.79), the remaining constants are obtained as: $c_1 = F_0$ and $c_2 = -F_0 L$. Thus, the distributions of deflection, rotational angle, bending moment, and shear force can be stated as:

$$u_z(x) = \frac{F_0 L^3}{EI_y} \left\{ \frac{1}{6}\left(\frac{x}{L}\right)^3 - \frac{1}{2}\left(\frac{x}{L}\right)^2 \right\}, \tag{7.5}$$

$$\varphi_y(x) = \frac{F_0 L^2}{EI_y} \left\{ -\frac{1}{2}\left(\frac{x}{L}\right)^2 + \left(\frac{x}{L}\right) \right\}, \tag{7.6}$$

$$M_y(x) = F_0 L \left\{ -\left(\frac{x}{L}\right) + 1 \right\}, \tag{7.7}$$

$$Q_z(x) = -F_0. \tag{7.8}$$

The other three cases can be solved in a similar way and the final results for the distributions are summarized in the following:

Case (b): Single moment M_0 at $x = L$

$$u_z(x) = \frac{M_0 L^2}{EI_y} \left\{ \frac{1}{2} \left(\frac{x}{L} \right)^2 \right\} , \tag{7.9}$$

$$\varphi_y(x) = -\frac{M_0 L}{EI_y} \left(\frac{x}{L} \right) , \tag{7.10}$$

$$M_y(x) = -M_0 , \tag{7.11}$$

$$Q_z(x) = 0 . \tag{7.12}$$

Case (c): Displacement u_0 at $x = L$

$$u_z(x) = \left\{ \frac{1}{2} \left(\frac{x}{L} \right)^3 - \frac{3}{2} \left(\frac{x}{L} \right)^2 \right\} u_0 , \tag{7.13}$$

$$\varphi_y(x) = \left\{ -\frac{3}{2} \left(\frac{x}{L} \right)^2 + 3 \left(\frac{x}{L} \right) \right\} \frac{u_0}{L} , \tag{7.14}$$

$$M_y(x) = \frac{3EI_y u_0}{L^2} \left\{ -\left(\frac{x}{L} \right) + 1 \right\} , \tag{7.15}$$

$$Q_z(x) = -\frac{3EI_y u_0}{L^3} . \tag{7.16}$$

Case (d): Rotation φ_0 at $x = L$

$$u_z(x) = \frac{\varphi_0 L}{2} \left(\frac{x}{L} \right)^2 , \tag{7.17}$$

$$\varphi_y(x) = -\varphi_0 \left(\frac{x}{L} \right) , \tag{7.18}$$

$$M_y(x) = -\frac{\varphi_0 EI_y}{L} , \tag{7.19}$$

$$Q_z(x) = 0 . \tag{7.20}$$

The constants of integration for all cases are summarized in Table 7.1.

2.2 Beam fixed at both ends: analytical solution of the deformations
The general solution is given by Eqs. (2.77)–(2.80):

Table 7.1 Constants of integration for the problems shown in Fig. 2.45

Case	c_1	c_2	c_3	c_4
(a)	F_0	$-F_0 L$	0	0
(b)	0	M_0	0	0
(c)	$\frac{3EI_y u_0}{L^3}$	$-\frac{3EI_y u_0}{L^2}$	0	0
(d)	0	$\frac{EI_y \varphi_0}{L}$	0	0

$$u_z(x) = \frac{1}{EI_y}\left(\frac{q_0 x^4}{24} + \frac{c_1 x^3}{6} + \frac{c_2 x^2}{2} + c_3 x + c_4\right), \tag{7.21}$$

$$Q_z(x) = -q_0 x - c_1, \tag{7.22}$$

$$M_y(x) = -\frac{q_0 x^2}{2} - c_1 x - c_2, \tag{7.23}$$

$$\varphi_y(x) = -\frac{du_z(x)}{dx} = -\frac{1}{EI_y}\left(\frac{q_0 x^3}{6} + \frac{c_1 x^2}{2} + c_2 x + c_3\right). \tag{7.24}$$

(a) Single force case $(0 \leq x \leq \frac{L}{2})$

Introducing the boundary conditions $u_z(0) = 0$, $\varphi_y(0) = 0$, $Q_z(0) = \frac{F_0}{2}$, $\varphi_y(\frac{L}{2}) = 0$ for the case $q_0 = 0$ in Eqs. (7.21)–(7.24) allows to determine the constant of integrations as:

$$c_1 = -\frac{F_0}{2}, \ c_2 = \frac{F_0 L}{8}, \ c_3 = 0, \ c_4 = 0. \tag{7.25}$$

Thus, the bending line can be stated as

$$u_z(x) = \frac{1}{EI_y}\left(-\frac{F_0 x^3}{12} + \frac{F_0 L x^2}{16}\right), \tag{7.26}$$

$$= \frac{F_0 L^3}{EI_y}\left(-\frac{1}{12}\left[\frac{x}{L}\right]^3 + \frac{1}{16}\left[\frac{x}{L}\right]^2\right), \tag{7.27}$$

or for the rotational field:

$$\varphi_y(x) = \frac{1}{EI_y}\left(\frac{F_0 x^2}{4} - \frac{F_0 L x}{8}\right), \tag{7.28}$$

$$= +\frac{F_0 L^2}{EI_y}\left(\frac{1}{4}\left[\frac{x}{L}\right]^2 - \frac{1}{8}\left[\frac{x}{L}\right]^1\right). \tag{7.29}$$

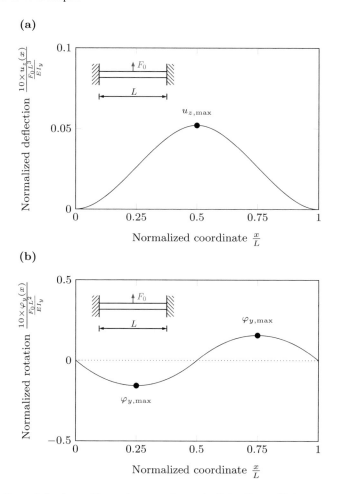

Fig. 7.3 a Beam deflection and **b** rotation according to the Euler–Bernoulli theory

The extremal values are obtained as follows:

$$u_{z,\text{max}} = \frac{F_0 L^3}{192 E I_y}, \; \varphi_{y,\text{max}} = \varphi_y(\tfrac{L}{4}) = -\frac{F_0 L^2}{64 E I_y}. \tag{7.30}$$

The graphical representation of the displacement and rotational field is provided in Fig. 7.3.

(b) Distributed load case

Introducing the boundary conditions $u_z(0) = 0$, $\varphi_y(0) = 0$, $Q_z(0) = \frac{q_0 L}{2}$, $\varphi_y(\tfrac{L}{2}) = 0$ in Eqs. (7.21)–(7.24) allows to determine the constant of integrations as:

$$c_1 = -\frac{q_0 L}{2}, \; c_2 = \frac{q_0 L^2}{12}, \; c_3 = 0, \; c_4 = 0. \tag{7.31}$$

Thus, the bending line can be stated as

$$u_z(x) = \frac{1}{EI_y}\left(\frac{q_0 x^4}{24} - \frac{q_0 L x^3}{12} + \frac{q_0 L^2 x^2}{24}\right), \tag{7.32}$$

$$= \frac{q_0 L^4}{EI_y}\left(\frac{1}{24}\left[\frac{x}{L}\right]^4 - \frac{1}{12}\left[\frac{x}{L}\right]^3 + \frac{1}{24}\left[\frac{x}{L}\right]^2\right), \tag{7.33}$$

or for the rotational field:

$$\varphi_y(x) = -\frac{1}{EI_y}\left(\frac{q_0 x^3}{6} - \frac{q_0 L x^2}{4} + \frac{q_0 L^2 x}{12}\right). \tag{7.34}$$

$$= +\frac{q_0 L^3}{EI_y}\left(-\frac{1}{6}\left[\frac{x}{L}\right]^3 + \frac{1}{4}\left[\frac{x}{L}\right]^2 - \frac{1}{12}\left[\frac{x}{L}\right]^1\right). \tag{7.35}$$

The extremal values are obtained as follows:

$$u_{z,\text{max}} = \frac{q_0 L^4}{384 EI_y}, \ \varphi_{y,\text{max}} = \varphi_y(\tfrac{3-\sqrt{3}}{6}L) = -\frac{\sqrt{3}q_0 L^3}{216 EI_y}. \tag{7.36}$$

The graphical representation of the displacement and rotational field is provided in Fig. 7.4.

2.3 Euler–Bernoulli beam with quadratic distributed load: analytical solution of the bending line

Four times integration of the fourth-order differential equation provided in Table 2.7 gives:

$$EI_y \frac{d^4 u_z(x)}{dx^4} = q_0\left(\frac{x^2}{L^2} - \frac{2x}{L} + 1\right), \tag{7.37}$$

$$EI_y \frac{d^3 u_z(x)}{dx^3} = -Q_z(x) = q_0\left(\frac{x^3}{3L^2} - \frac{x^2}{L} + x\right) + c_1, \tag{7.38}$$

$$EI_y \frac{d^2 u_z(x)}{dx^2} = -M_y(x) = q_0\left(\frac{x^4}{12L^2} - \frac{x^3}{3L} + \frac{x^2}{2}\right) + c_1 x + c_2, \tag{7.39}$$

$$EI_y \frac{d^1 u_z(x)}{dx^1} = -EI_y \varphi_y(x) = q_0\left(\frac{x^5}{60L^2} - \frac{x^4}{12L} + \frac{x^3}{6}\right) + \frac{c_1 x^2}{2} + c_2 x + c_3, \tag{7.40}$$

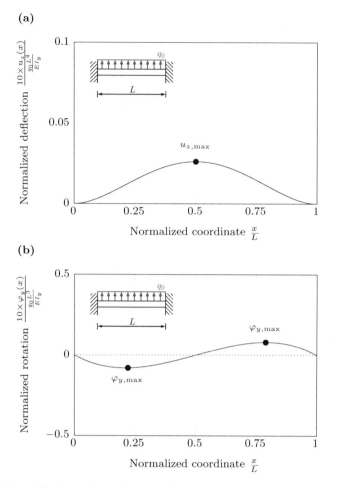

Fig. 7.4 **a** Beam deflection and **b** rotation according to the Euler–Bernoulli theory

$$EI_y u_z(x) = q_0 \left(\frac{x^6}{360L^2} - \frac{x^5}{60L} + \frac{x^4}{24} \right) + \frac{c_1 x^3}{6} + \frac{c_2 x^2}{2} + c_3 x + c_4 . \quad (7.41)$$

The consideration of the boundary conditions, i.e., $u_z(0) = 0$, $M_y(0) = 0$, $u_z(L) = 0$, and $\varphi_y(L) = 0$, allows to determine the constants of integration as follows:

$$c_2 = 0, \qquad\qquad\qquad c_4 = 0, \qquad\qquad (7.42)$$

$$c_1 = -\frac{13}{60} q_0 L, \qquad\qquad c_3 = \frac{1}{120} q_0 L^3 . \qquad (7.43)$$

Fig. 7.5 Beam deflection
along the major axis

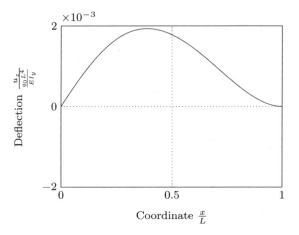

Thus, we can state the bending line as:

$$u_z(x) = \frac{q_0 L^4}{EI_y} \left[\frac{1}{360} \left(\frac{x}{L}\right)^6 - \frac{1}{60} \left(\frac{x}{L}\right)^5 + \frac{1}{24} \left(\frac{x}{L}\right)^4 - \frac{13}{360} \left(\frac{x}{L}\right)^3 + \frac{1}{120} \left(\frac{x}{L}\right)^1 \right].$$

(7.44)

The normalized deflection is shown in Fig. 7.5.

In the same approach, the internal bending moment distribution is obtained as (see Fig. 7.6):

$$M_y(x) = -q_0 L^2 \left[\frac{1}{12} \left(\frac{x}{L}\right)^4 - \frac{1}{3} \left(\frac{x}{L}\right)^3 + \frac{1}{2} \left(\frac{x}{L}\right)^2 - \frac{13}{60} \left(\frac{x}{L}\right)^1 \right].$$

(7.45)

2.4 Unsymmetrical bending of a Z-profile

Let us split the Z-profile into three simple shapes, i.e., three rectangles, as shown in Fig. 7.7. In addition, an initial y'-z' coordinate system is introduced with its origin in the lower right corner, see Fig. 7.7a.

The coordinates of the single centroids as wells as the corresponding surface areas are summarized in Table 7.2.

The coordinates (y'_c, z'_c) of the centroid can be obtained from Eqs. (B.3) and (B.4) as:

$$y'_c = \frac{y'_1 A_1 + y'_2 A_2 + y'_3 A_3}{A_1 + A_2 + A_3} = 10.0 \, \text{mm},$$

(7.46)

$$z'_c = \frac{z'_1 A_1 + z'_2 A_2 + z'_3 A_3}{A_1 + A_2 + A_3} = 75.0 \, \text{mm}.$$

(7.47)

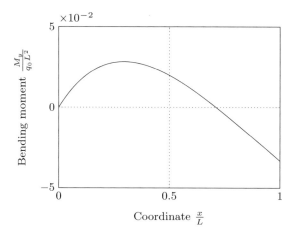

Fig. 7.6 Internal bending moment distribution along the major axis

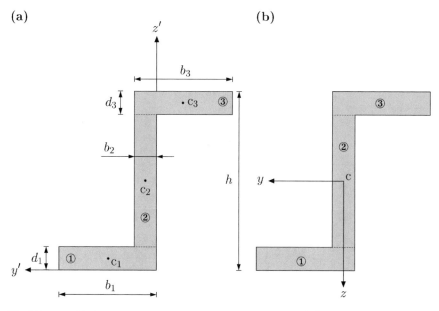

Fig. 7.7 Z-profile for unsymmetrical bending problem: **a** Initial y'-z' coordinate system for the determination of the centroid. **b** Cartesian y-z coordinate system whereas the origin coincides with the centroid

Table 7.2 Position of centroids in the y'-z' coordinate system and corresponding surface areas

Surface i	y_i'	z_i'	A_i
1	$\dfrac{b_1}{2}$	$\dfrac{d_1}{2}$	$b_1 \times d_1$
2	$\dfrac{b_2}{2}$	$d_1 + \dfrac{h - d_1 - d_3}{2}$	$b_2 \times (h - d_1 - d_3)$
3	$-\left(\dfrac{b_3}{2} - b_2\right)$	$h - \dfrac{d_3}{2}$	$b_3 \times d_3$

The next step is to calculate the second moments of area in the y-z coordinate system (see Fig. 7.7a). Based on the second moments for simple shapes provided in Tables 2.2 and 2.6 and the parallel axis theorem provided in Sect. B.3, the following relationships are obtained:

$$I_y = I_{y,1} + I_{y,2} + I_{y,3} \tag{7.48}$$

$$= \frac{1}{12}b_1 d_1^3 + (z_c - d_1/2)^2 A_1 + \frac{1}{12}b_2(h - d_1 - d_3)^3$$

$$+ \frac{1}{12}b_3 d_3^3 + (-z_c + d_3/2)^2 A_3 \tag{7.49}$$

$$= 1.7548 \times 10^7 \, \text{mm}^4 , \tag{7.50}$$

$$I_z = I_{z,1} + I_{z,2} + I_{z,3} \tag{7.51}$$

$$= \frac{1}{12}d_1 b_1^3 + (b_1/2 - y_c)^2 A_1 + \frac{1}{12}(h - d_1 - d_3)b_2^3$$

$$+ \frac{1}{12}d_3 b_3^3 + (-b_3/2 + b_2/2)^2 A_3 \tag{7.52}$$

$$= 6.9133 \times 10^6 \, \text{mm}^4 , \tag{7.53}$$

$$I_{yz} = I_{yz,1} + I_{yz,2} + I_{yz,3} \tag{7.54}$$

$$= -(z_c - d_1/2)(b_1/2 - y_c)A_1 - (0)(0)A_2$$

$$- (-z_c + d_3/2)(-b_3/2 + b_2/2)A_3 \tag{7.55}$$

$$= -8.1900 \times 10^6 \, \text{mm}^4 . \tag{7.56}$$

The principal moments of area in the η-ζ coordinate system result from Eqs. (2.103) and (2.104) as:

Fig. 7.8 Principal axis system η-ζ and orientation of the moment vector M

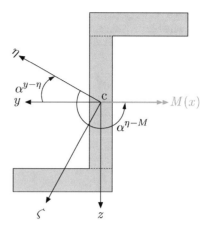

$$I_\eta = \frac{I_y + I_z}{2} + \sqrt{\left(\frac{I_y - I_z}{2}\right)^2 + I_{yz}^2} = 2.1996 \times 10^7 \, \text{mm}^4 \,, \qquad (7.57)$$

$$I_\zeta = \frac{I_y + I_z}{2} - \sqrt{\left(\frac{I_y - I_z}{2}\right)^2 + I_{yz}^2} = 2.4660 \times 10^6 \, \text{mm}^4 \,. \qquad (7.58)$$

The angle of rotation can be calculated based on the second moments of area in the x-y system as (see Eq. (2.105)):

$$\alpha^{y-\eta,\zeta} = \frac{1}{2} \times \arctan\left(\frac{2I_{yz}}{I_y - I_z}\right) = -0.497468 = -28.50° \,, \qquad (7.59)$$

whereas the second-order derivative of Eq. (2.106), i.e.,

$$\frac{d^2 I_\eta (\alpha^{y-\eta,\zeta})}{d(\alpha^{y-\eta,\zeta})^2} = -2\left(I_y - I_z\right)\cos(2\alpha^{y-\eta,\zeta}) - 4I_{yz}\sin(2\alpha^{y-\eta,\zeta})\,, \qquad (7.60)$$

$$= -3.91 \times 10^7 \, \text{mm}^4 < 0 \,, \qquad (7.61)$$

allows the identification $\alpha^{y-\eta,\zeta} = \alpha^{y-\eta}$ (see Fig. 7.8). The moment vector M must act in direction of a line which is perpendicular to the line of action of the external force F_0 (see Fig. 2.48). Based on the beam deformation, the moment vector must be introduced as shown in Fig. 7.8. The angle between the η-axis and the internal moment vector M can be calculated as follows:

$$\alpha^{\eta-M} = |\alpha^{y-\eta}| + 180° = 208.5° \,. \qquad (7.62)$$

Fig. 7.9 Orientation of the neutral fiber (n-n) and the critical stress points A and B

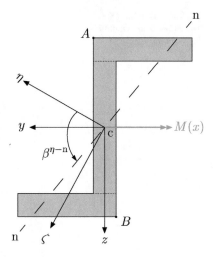

The rotational angle $\beta^{\eta-n}$ between the η-axis and the neutral fiber can be obtained from Eqs. (2.113) and (2.114)-(2.115) as (see Fig. 7.9):

$$\beta^{\eta-n} = \arctan\left(\frac{I_\eta}{I_\zeta} \times \frac{|M|\sin(\alpha^{\eta-M})}{|M|\cos(\alpha^{\eta-M})}\right) = 1.367195 = 78.33°\,. \tag{7.63}$$

From Fig. 7.9, it is possible to identify on both sides of the neutral fiber the critical stress points, i.e. points A and B. The stress calculation according to Eq. (2.116) requires to express the coordinates of these points in the η-ζ coordinate system. Using Eqs. (2.117) and (2.118), one can state:

$$\eta_A = y_A \cos(-\alpha^{y-\eta}) - z_A \sin(-\alpha^{y-\eta}) = 44.58\,\text{mm}\,, \tag{7.64}$$

$$\zeta_A = y_A \sin(-\alpha^{y-\eta}) + z_A \cos(-\alpha^{y-\eta}) = -61.14\,\text{mm}\,, \tag{7.65}$$

$$\eta_B = y_B \cos(-\alpha^{y-\eta}) - z_B \sin(-\alpha^{y-\eta}) = -44.58\,\text{mm}\,, \tag{7.66}$$

$$\zeta_B = y_B \sin(-\alpha^{y-\eta}) + z_B \cos(-\alpha^{y-\eta}) = 61.14\,\text{mm}\,. \tag{7.67}$$

Thus, the stresses in the critical points can be obtained as:

$$\sigma_{x,A} = |M|\left(\frac{\cos(\alpha^{\eta-M})}{I_\eta}\zeta_A - \frac{\sin(\alpha^{\eta-M})}{I_\zeta}\eta_A\right) = 92.98\,\text{MPa}\,, \tag{7.68}$$

$$\sigma_{x,B} = |M|\left(\frac{\cos(\alpha^{\eta-M})}{I_\eta}\zeta_B - \frac{\sin(\alpha^{\eta-M})}{I_\zeta}\eta_B\right) = -92.98\,\text{MPa}\,. \tag{7.69}$$

Fig. 7.10 Circular cross
section used for the
derivation of the shear stress
$\tau_{xz}(z)$

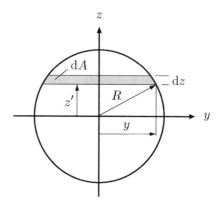

2.5 Calculation of the shear stress distribution in a circular cross section

Starting point is again the infinitesimal beam element given in Fig. 2.33. However,
the cross section is now as indicated in Fig. 7.10.

Force equilibrium in the x-direction gives:

$$\int \sigma_x(x)\, dA - \int \left(\sigma_x(x) + \frac{d\sigma_x(x)}{dx} dx \right) dA + \tau_{xz} \underbrace{2y}_{b(z)}\, dx = 0. \qquad (7.70)$$

It results from Eq. (2.66) after differentiation with respect to the x-coordinate:

$$\frac{d\sigma_x(x)}{dx} = +\frac{z}{I_y}\frac{dM_y(x)}{dx} \overset{(2.75)}{=} \frac{Q_z(x) \times z}{I_y}. \qquad (7.71)$$

Thus,

$$\tau_{xy} = \frac{Q_z(x)}{2yI_y} \int z'\, dA = \frac{Q_z(x)}{2yI_y} \mathcal{H}_y(z). \qquad (7.72)$$

Considering that $dA = 2y dz$ and $y = \sqrt{R^2 - z^2}$, the first moment of area for the
part of the cross section with $z' \le z \le R$ is obtained as:

$$\mathcal{H}_y(z) = \int_z^R z'2y dz' = \int_z^R z'2\sqrt{R^2 - z'^2}dz' = \frac{2}{3}\left(R^2 - z^2\right)^{3/2}. \qquad (7.73)$$

The final shear stress distribution is then obtained as:

$$\tau_{xy} = \frac{Q_z(x)}{3I_y}\left(R^2 - z^2\right) = \frac{4Q_z(x)}{3\pi R^2}\left(1 - \left(\frac{z}{R}\right)^2\right). \qquad (7.74)$$

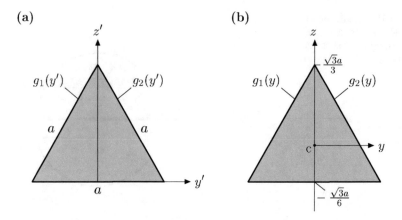

Fig. 7.11 Triangular section: **a** Initial y'-z' coordinate system for the determination of the centroid. **b** Cartesian y-z coordinate system whereas the origin coincides with the centroid

Maximum shear stress for $z = 0$:

$$\tau_{xz,\max} = \frac{Q_z(x)R^2}{3I_y} = \frac{4Q_z(x)}{3\pi R^2} = \frac{4Q_z(x)}{3A}. \tag{7.75}$$

2.6 Calculation of the shear stress distribution in a triangular cross section

The first step of the solution procedure is to calculate the centroid of the equilateral triangle. To this end, the initial y'-z' Cartesian coordinate system is introduced, see Fig. 7.11a. Since we have a singly symmetric cross section, the centroid must be on the symmetry line $y' = 0$ and it remains to determine the vertical location of the centroid.

Based on Eq. (B.1), the vertical coordinate of the centroid in the y'-z' system can be expressed as:

$$z_c' = \frac{\int z' \, dA'}{\int dA'} = \frac{\displaystyle\int_{y'=-\frac{a}{2}}^{y'=0} \int_{z'=0}^{z'=g_1(y')} z' dz' dy' + \int_{y'=0}^{y'=\frac{a}{2}} \int_{z'=0}^{z'=g_2(y')} z' dz' dy'}{\displaystyle\int_{y'=-\frac{a}{2}}^{y'=0} \int_{z'=0}^{z'=g_1(y')} dz' dy' + \int_{y'=0}^{y'=\frac{a}{2}} \int_{z'=0}^{z'=g_2(y')} dz' dy'} \tag{7.76}$$

$$= \frac{\frac{a^3}{8}}{\frac{\sqrt{3}}{4} a^2} = \frac{\sqrt{3}}{6} a, \tag{7.77}$$

where the following functions of the inclined sides were used:

Fig. 7.12 Reference section
(marked in gray) for the
calculation of the first
moment of area $\mathcal{H}_y(z)$

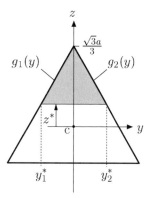

$$g_1(y') = \frac{\sqrt{3}}{2}a + \sqrt{3}y', \qquad (7.78)$$

$$g_2(y') = \frac{\sqrt{3}}{2}a - \sqrt{3}y'. \qquad (7.79)$$

Thus, the coordinates of the centroid in the y'-z' coordinate system are obtained as $(y'_c, z'_c) = (0, \frac{\sqrt{3}}{6}a)$. The next step is to move the initial coordinate system into the centroid of the section, see Fig. 7.11b. This new system is now called y-z. The functions of the inclined sides can be expressed now as:

$$g_1(y) = \frac{\sqrt{3}}{3}a + \sqrt{3}y, \qquad (7.80)$$

$$g_2(y) = \frac{\sqrt{3}}{3}a - \sqrt{3}y. \qquad (7.81)$$

The computation of the shear stress according to Eq. (2.127) requires the evaluation of the second moment of area I_y and the first moment of area $\mathcal{H}_y(z)$. Let us focus first on the second moment of area. According to Eq. (B.5), we get the following expression:

$$I_y = \int_A z^2 \mathrm{d}A = \int_{y=-\frac{a}{2}}^{y=0} \int_{z=-\frac{\sqrt{3}}{6}a}^{z=g_1(y)} z^2 \mathrm{d}z\mathrm{d}y + \int_{y=0}^{y=\frac{a}{2}} \int_{z=-\frac{\sqrt{3}}{6}a}^{z=g_2(y)} z^2 \mathrm{d}z\mathrm{d}y \qquad (7.82)$$

$$= \frac{\sqrt{3}}{96}a^4. \qquad (7.83)$$

The next step is to calculate the first moment of area $\mathcal{H}_y(z)$ for the part of the cross section shown in Fig. 7.12.

According to Eq. (2.127), one can state:

$$\mathcal{H}_y(z) = \int z \, \mathrm{d}A \tag{7.84}$$

$$= \int\limits_{y=y_1^*}^{y=0} \int\limits_{z=z^*}^{z=g_1(y)} z \, \mathrm{d}z \mathrm{d}y + \int\limits_{y=0}^{y=y_2^*} \int\limits_{z=z^*}^{z=g_2(y)} z \, \mathrm{d}z \mathrm{d}y \tag{7.85}$$

$$= \int\limits_{y=-\frac{a}{3}+\frac{z^*}{\sqrt{3}}}^{y=0} \int\limits_{z=z^*}^{z=g_1(y)} z \, \mathrm{d}z \mathrm{d}y + \int\limits_{y=0}^{y=\frac{a}{3}-\frac{z^*}{\sqrt{3}}} \int\limits_{z=z^*}^{z=g_2(y)} z \, \mathrm{d}z \mathrm{d}y \tag{7.86}$$

$$= \frac{a^3}{27} - \frac{az^2}{3} + \frac{2z^3}{3\sqrt{3}} = \underbrace{\left(\frac{a}{3} - \frac{z}{\sqrt{3}}\right)^2}_{\frac{b(z^*)}{2}} \left(\frac{2z}{\sqrt{3}} + \frac{a}{3}\right). \tag{7.87}$$

Thus, the shear stress distribution can be stated as follows:

$$\tau_{xz}(z) = \frac{Q_z(x)\mathcal{H}_y(z)}{I_y b(z)} \tag{7.88}$$

$$= \frac{Q_z(x)}{2I_y} \times \frac{\left(\frac{a}{3} - \frac{z}{\sqrt{3}}\right)^2 \left(\frac{2z}{\sqrt{3}} + \frac{a}{3}\right)}{\left(\frac{a}{3} - \frac{z}{\sqrt{3}}\right)} \tag{7.89}$$

$$= \frac{Q_z(x)}{2I_y} \times \left(\frac{a}{3} - \frac{z}{\sqrt{3}}\right)\left(\frac{2z}{\sqrt{3}} + \frac{a}{3}\right). \tag{7.90}$$

Equation (7.90) fulfills the boundary condition that the shear stress must reduce to zero at the top and bottom surface, i.e. $\tau_{xz}(z = \frac{\sqrt{3}a}{3}) = 0$ and $\tau_{xz}(z = -\frac{\sqrt{3}a}{6}) = 0$. The shear stress in the neutral fiber, i.e., for $z = 0$, is obtained as:

$$\tau_{xz}(z = 0) = \frac{Q_z(x)a^2}{18I_y} = \frac{16Q_z(x)}{3\sqrt{3}a^2}. \tag{7.91}$$

However, the stress value in the neutral fiber (see Eq. (7.91)) is not the maximum value. The condition

$$\frac{\partial \tau_{xz}(z)}{\partial z} \overset{!}{=} 0 = \frac{2a}{3\sqrt{3}} - \frac{a}{3\sqrt{3}} - \frac{4z}{3} \tag{7.92}$$

Fig. 7.13 Configuration for
the calculation of the shear
stress distribution:
continuous ring cross section

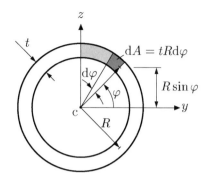

provides the location of the maximum as:

$$z_{max} = \frac{\sqrt{3}a}{12},$$ (7.93)

and the maximum shear stress is obtained as follows:

$$\tau_{xz}(z_{max}) = \frac{Q_z(x)a^2}{16I_y} = \frac{2\sqrt{3}Q_z(x)}{a^2}.$$ (7.94)

2.7 Calculation of the shear stress distributions in circular ring cross sections

It is possible to simply adopt Eq. (2.140) to the case of a thin circular ring cross
section (thickness t) as follows:

$$\tau = \frac{Q_z(x)\mathcal{H}_y}{I_y t}.$$ (7.95)

The second moment of area I_y is for both configurations the same and can be calculated based on Eq. (B.5):

$$I_y = \int_A z^2 dA = \int_0^{2\pi} R^2 \sin^2(\varphi) t\, Rd\varphi = R^3 t \int_0^{2\pi} 2\sin^2(\varphi)d\varphi = \pi R^3 t.$$ (7.96)

Thus, it remains to calculate the first moments of area $\mathcal{H}_y$ for both configurations.
Let us look first at the continuous ring cross section as shown in Fig. 7.13.
    The first moment reads[1]:

---

[1] It should be noted that an integration over the boundaries φ and $\pi - \varphi$ would result in twice the
value. However, this would not consider that the shear stress is distributed in the left- and right-hand
part of the ring.

Fig. 7.14 Shear stress distribution for the continuous ring cross section

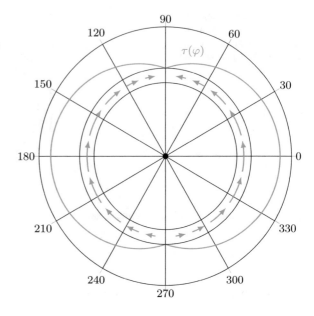

$$\mathcal{H}_y = \int z dA = \int\limits_{\varphi}^{\frac{\pi}{2}} R\sin(\varphi')t\,Rd\varphi' = R^2 t \int\limits_{\varphi}^{\frac{\pi}{2}} \sin(\varphi')d\varphi' = R^2 t \cos(\varphi),\quad (7.97)$$

and the shear stress distribution can be expressed as:

$$\tau(\varphi) = \frac{Q_z(x)\mathcal{H}_y}{I_y t} = \frac{Q_z(x)\cos(\varphi)}{\pi R t}. \qquad (7.98)$$

It can be seen from Fig. 7.14. That the shear stress is zero for $\varphi = 90°$ and $270°$ and the maxima are reached for $0°$ and $180°$. It should be noted here that for this doubly symmetric cross section the shear center is located at the origin of the ring, i.e. ($y_{sc} = 0$, $z_{sc} = 0$).

Let us look now at the slotted ring cross section as shown in Fig. 7.15. The first moment reads for the slotted ring:

$$\mathcal{H}_y = \int z dA = \int\limits_{0}^{\varphi} R\sin(\varphi')t\,Rd\varphi' = R^2 t \int\limits_{0}^{\varphi} \sin(\varphi')d\varphi' = R^2 t(1 - \cos(\varphi)),$$

$$(7.99)$$

and the shear stress distribution can be expressed as:

Fig. 7.15 Configuration for the calculation of the shear stress distribution: slotted ring cross section

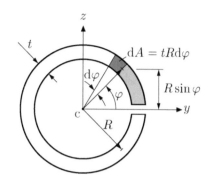

Fig. 7.16 Shear stress distribution for the slotted ring cross section

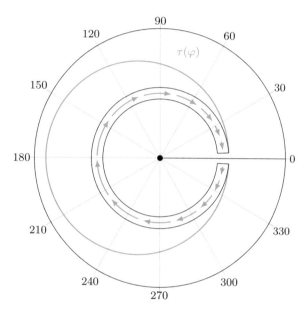

$$\tau(\varphi) = \frac{Q_z(x)\mathcal{H}_y}{I_y t} = \frac{Q_z(x)(1 - \cos(\varphi))}{\pi R t}. \tag{7.100}$$

It can be seen from Fig. 7.16. That the shear stress is zero for $\varphi = 0°$ and $360°$ and the maximum is reached for $180°$. The location of the shear center for the slotted ring can be calculated by considering a similar situation as shown in Figs. 2.42b and 2.43. The moment equilibrium with respect to the centroid gives:

$$\sum_{i=1}^{2} M_i = 0 \quad \Leftrightarrow \quad -F_z|y_{sc}| + \int_0^{2\pi} R\tau(\varphi)dA = 0, \tag{7.101}$$

or rearranged for the location of the shear center[2]:

$$|y_{sc}| = \frac{R \int\limits_0^{2\pi} \frac{F_z(1-\cos(\varphi))}{\pi R t} \mathrm{d}A}{F_z} = \frac{R}{\pi}[\varphi - \sin(\varphi)]_0^{2\pi} = 2R \,. \qquad (7.102)$$

7.2   Problems from Chap. 3

3.1 Cantilever Timoshenko beam with different end loads and deformations

The free-body diagrams look as shown in Fig. 7.1 since this is independent of the applied beam theory. The same holds for the reactions at the support, see Eqs. (7.1) and (7.2).

Case (a): Single force F_0 at $x = L$

The boundary conditions remain as given in Eqs. (7.3) and (7.4), i.e.,

$$u_z(0) = 0 \,, \qquad \varphi_y(0) = 0 \,, \qquad Q_z(L) = -F_0 \,, \qquad M_y(L) = 0 \,. \qquad (7.103)$$

Consideration of boundary condition $(7.103)_1$ in Eq. (3.93) gives $c_4 = 0$. In a similar way, the first constant of integration can be obtained by considering the boundary condition $(7.103)_3$ in the general expression for the shear force distribution (3.96): $c_1 = F_0$. Introducing the boundary conditions at the left-hand end for the rotation and at the right-hand end for the bending moment, i.e. Eq. $(7.103)_2$ and $(7.103)_4$ in the expressions for the rotation and the bending moment according to Eqs. (3.94) and (3.95), the remaining constants are obtained as: $c_3 = -\frac{EI_y F_0}{k_s AG}$ and $c_2 = -F_0 L$. Thus, the distributions of deflection, rotational angle, bending moment, and shear force can be stated as:

$$u_z(x) = \underbrace{\frac{F_0 L^3}{EI_y}\left\{\frac{1}{6}\left(\frac{x}{L}\right)^3 - \frac{1}{2}\left(\frac{x}{L}\right)^2\right\} - \frac{F_0 L}{k_s AG}\left(\frac{x}{L}\right)}_{\text{EB solution}}, \qquad (7.104)$$

$$\varphi_y(x) = \underbrace{\frac{F_0 L^2}{EI_y}\left\{-\frac{1}{2}\left(\frac{x}{L}\right)^2 + \left(\frac{x}{L}\right)\right\}}_{\text{EB solution}}, \qquad (7.105)$$

---

[2]The shear center is located on the left-hand side of the centroid in this case.

$$M_y(x) = F_0 L \underbrace{\left\{ -\left(\frac{x}{L}\right) + 1 \right\}}_{\text{EB solution}}, \tag{7.106}$$

$$Q_z(x) = \underbrace{-F_0}_{\text{EB solution}} . \tag{7.107}$$

The other three cases can be solved in a similar way and the final results for the distributions are summarized in the following:

Case (b): Single moment M_0 at $x = L$

$$u_z(x) = \underbrace{\frac{M_0 L^2}{E I_y} \left\{ \frac{1}{2} \left(\frac{x}{L}\right)^2 \right\}}_{\text{EB solution}}, \tag{7.108}$$

$$\varphi_y(x) = \underbrace{-\frac{M_0 L}{E I_y} \left(\frac{x}{L}\right)}_{\text{EB solution}}, \tag{7.109}$$

$$M_y(x) = \underbrace{-M_0}_{\text{EB solution}}, \tag{7.110}$$

$$Q_z(x) = \underbrace{0}_{\text{EB solution}} . \tag{7.111}$$

Case (c): Displacement u_0 at $x = L$

$$u_z(x) = \left\{ \frac{1}{2} \times \frac{1}{1 + \frac{3EI_y}{k_s AGL^2}} \left(\frac{x}{L}\right)^3 - \frac{3}{2} \times \frac{1}{1 + \frac{3EI_y}{k_s AGL^2}} \left(\frac{x}{L}\right)^2 \right.$$
$$\left. - \frac{1}{1 + \frac{k_s AGL^2}{3EI_y}} \left(\frac{x}{L}\right) \right\} u_0, \tag{7.112}$$

$$\varphi_y(x) = \left\{ -\frac{3}{2} \times \frac{1}{1 + \frac{3EI_y}{k_s AGL^2}} \left(\frac{x}{L}\right)^2 + 3 \times \frac{1}{1 + \frac{3EI_y}{k_s AGL^2}} \left(\frac{x}{L}\right) \right\} \frac{u_0}{L}, \tag{7.113}$$

$$M_y(x) = \frac{3EI_y u_0}{L^2} \left\{ -\frac{1}{1 + \frac{3EI_y}{k_s AGL^2}} \left(\frac{x}{L}\right) + \frac{1}{1 + \frac{3EI_y}{k_s AGL^2}} \right\}, \tag{7.114}$$

$$Q_z(x) = -\frac{3EI_y u_0}{L^3 + \frac{3EI_y L}{k_s AG}} . \tag{7.115}$$

Table 7.3 Constants of integration for the problems shown in Fig. 3.17

Case	c_1	c_2	c_3	c_4
(a)	F_0	$-F_0 L$	$-\dfrac{EI_y F_0}{k_s AG}$	0
(b)	0	M_0	0	0
(c)	$\dfrac{3EI_y u_0}{L^3 + \dfrac{3EI_y L}{k_s AG}}$	$-\dfrac{3EI_y u_0}{L^2 + \dfrac{3EI_y}{k_s AG}}$	$-\dfrac{EI_y u_0}{L + \dfrac{k_s AGL^3}{3EI_y}}$	0
(d)	0	$\dfrac{EI_y \varphi_0}{L}$	0	0

It should be noted here that the expression (7.112)–(7.115) reduce for $k_s AG \to \infty$ to the classical Euler–Bernoulli solutions given in Eqs. (7.13)–(7.16).

Case (d): Rotation φ_0 at $x = L$

$$u_z(x) = \underbrace{\frac{\varphi_0 L}{2}\left(\frac{x}{L}\right)^2}_{\text{EB solution}}, \tag{7.116}$$

$$\varphi_y(x) = \underbrace{-\varphi_0\left(\frac{x}{L}\right)}_{\text{EB solution}}, \tag{7.117}$$

$$M_y(x) = \underbrace{-\frac{\varphi_0 EI_y}{L}}_{\text{EB solution}}, \tag{7.118}$$

$$Q_z(x) = \underbrace{0}_{\text{EB solution}}. \tag{7.119}$$

The constants of integration for all cases are summarized in Table 7.3.

3.2 Calculation of the shear correction factor for a rectangular cross section

The total shear strain energy for a pure shear stress state (just τ_{xz} acting) can be stated for an infinitesimal volume element (dc) as follows [2]:

$$d\Pi = \tau d\gamma d\Omega = G\gamma d\gamma d\Omega = \frac{G\gamma^2}{2}d\Omega = \frac{\tau^2}{2G}d\Omega . \tag{7.120}$$

The equivalence between the energies yields:

$$\int_\Omega \frac{1}{2G}\tau_{xz}^2 d\Omega \overset{!}{=} \int_{\Omega_s} \frac{1}{2G}\left(\frac{Q_z}{A_s}\right)^2 d\Omega_s , \tag{7.121}$$

Table 7.4 Constants of integration for the problems shown in Fig. 4.8

Case	c_1	c_2	c_3	c_4
(a)	F_0	$-F_0 L$	$-\dfrac{EI_y F_0}{\frac{2}{3} AG}$	0
(b)	0	M_0	0	0
(c)	$\dfrac{3EI_y u_0}{L^3 + \frac{3EI_y L}{\frac{2}{3} AG}}$	$-\dfrac{3EI_y u_0}{L^2 + \frac{3EI_y}{\frac{2}{3} AG}}$	$-\dfrac{EI_y u_0}{L + \frac{\frac{2}{3} AGL^3}{3EI_y}}$	0
(d)	0	$\dfrac{EI_y \varphi_0}{L}$	0	0

$$k_s = \frac{Q_z}{A \int_A \tau_{xz}^2 \, dA} = \frac{5}{6}. \tag{7.122}$$

7.3 Problems from Chap. 4

4.1 Cantilever Levinson beam with different end loads and deformations

The solution procedure follows exactly Prob. 3.1 and the obtained results can be adjusted to the Levinson problem by the substitution $k_s \to \frac{2}{3}$. The constants of integration for all cases are summarized in Table 7.4.

4.2 Cantilever Levinson beam with point load: effective stress over cross section

For the beam configuration according to Fig. 4.9, the internal bending moment and shear force distribution results in:

$$M_y(x) = F_0(x - L), \tag{7.123}$$
$$Q_z(x) = -F_0. \tag{7.124}$$

Assuming the von Mises stress hypothesis according to Eq. (4.82) (see [1] for details) and the expressions for the normal and shear stress distributions[3] according to Eqs. (2.66) and (2.131), the effective stress can be stated as follows:

$$\sigma_{\text{eff}} = \sqrt{\sigma^2 + 3\tau^2} = \sqrt{\left(\frac{M_y}{I_y} z\right)^2 + 3\left(\frac{3Q_z}{2A} \frac{h^2 - 4z^2}{h^2}\right)^2} \tag{7.125}$$

$$= \sqrt{\left(\frac{F_0(x - L)}{I_y} z\right)^2 + 3\left(-\frac{3F_0}{2A} \frac{h^2 - 4z^2}{h^2}\right)^2}. \tag{7.126}$$

---

[3] These equations from the Euler–Bernoulli beam chapter are still valid for the Levinson beam.

Fig. 7.17 Effective stress
for a Levinson beam for
different slenderness ratios:
a slender with $\frac{h}{L} = 0.1$, **b**
compact with $\frac{h}{L} = 0.6$; **c**
compact with $\frac{h}{L} = 0.9$

(a)

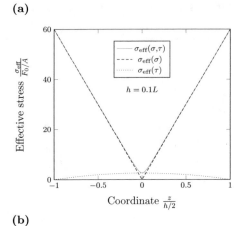

(b)

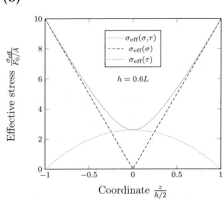

(c)

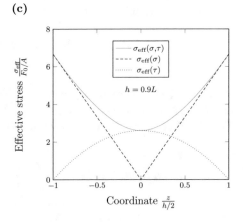

It follows from the above equation that the maximum occurs at the clamping point, i.e. $x = 0$. If we continue to consider a square cross-section with a side length h, one gets $I_y = \frac{h^4}{12} = \frac{Ah^2}{12}$. Thus, the effective stress results to:

$$\sigma_{\text{eff}} = \frac{F_0}{A}\sqrt{\left(\frac{12Lz}{h^2}\right)^2 + 3\left(\frac{3(h^2 - 4z^2)}{2h^2}\right)^2},\qquad(7.127)$$

or in a normalized representation:

$$\frac{\sigma_{\text{eff}}}{\frac{F_0}{A}} = \sqrt{\left(\frac{6L}{h}\left[\frac{z}{\frac{h}{2}}\right]\right)^2 + 3\left(\frac{3}{2}\left(1 - \left[\frac{z}{\frac{h}{2}}\right]^2\right)\right)^2}.\qquad(7.128)$$

The evaluation of this relation for different height to length ratios is shown in Fig. 7.17. It can be seen that as the beam length reduces, the significance of the shear stress increases in relation to the normal stress. Although the shear stress distribution remains unchanged, the magnitude of the normal stress (shorter lever to the clamping point) is reduced. The most important conclusion, however, is that the maximum of the effective stress occurs at the top or bottom side of the beam.

7.4 Problems from Chap. 5

5.1 Maximum beam deflections according to the Levinson and Timoshenko theory for different Poisson's ratios

The investigation of the influence of Poisson's ratio on the normalized maximum deflections can be based on the equations provided in Tables 5.1–5.3. The graphical evaluation for $\nu = 0.0, 0.3$, and 0.5 is shown in Figs. 7.19 and 7.18. It can be seen that an increasing Poisson's ratio leads to an increased maximum deflection.

5.2 Maximum beam deflections according to the Timoshenko theory for beams with circular cross sections

The solution for case (a) can be based on Eq. (5.5) and subsequent consideration of the circular cross section by setting $k_s = \frac{9}{10}$ and $I_y = \frac{\pi d^4}{64} = \frac{Ad^2}{16}$:

$$\frac{u_z^{\text{T}}(L)}{\frac{F_0 L^3}{3EI_y}} = 1 + \frac{5(1 + \nu)}{12}\left(\frac{h}{L}\right)^2.\qquad(7.129)$$

Fig. 7.18 Maximum beam deflections according to the Timoshenko theory for different Poisson's ratios: **a** cantilever configuration with point load, **b** cantilever configuration with distributed load, and **c** simply supported with point load

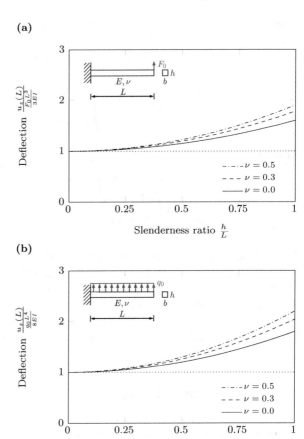

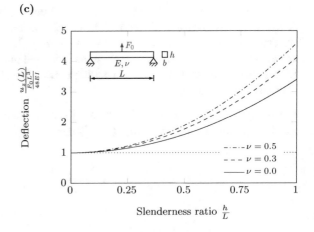

Fig. 7.19 Maximum beam deflections according to the Levinson theory for different Poisson's ratios: **a** cantilever configuration with point load, **b** cantilever configuration with distributed load, and **c** simply supported with point load

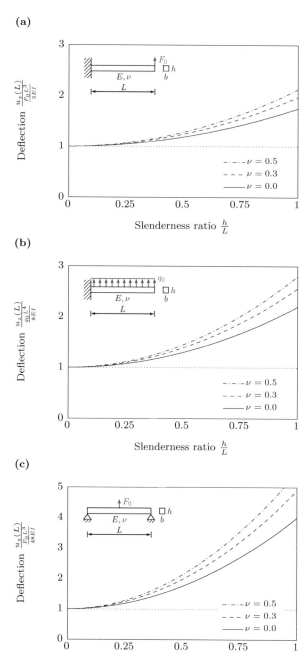

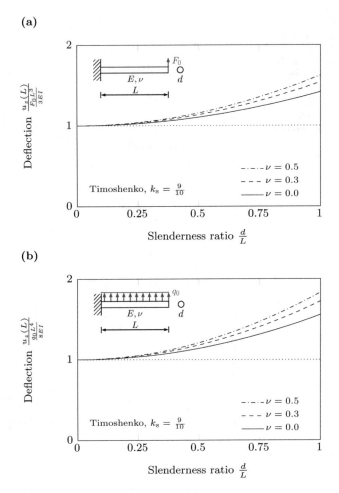

Fig. 7.20 Maximum deflections according to the Timoshenko theory for different Poisson's ratios of a circular cross section beam: **a** cantilever configuration with point load, **b** cantilever configuration with distributed load

Following the same approach, the solution for case (b) can be based on Eq. (5.11) and subsequent consideration of the circular cross section by setting $k_s = \frac{9}{10}$ and $I_y = \frac{\pi d^4}{64} = \frac{A d^2}{16}$:

$$\frac{u_z^{\mathrm{T}}(L)}{\frac{q_0 L^4}{8 E I_y}} = 1 + \frac{5(1+\nu)}{9}\left(\frac{h}{L}\right)^2. \tag{7.130}$$

The graphical evaluation of Eqs. (7.129) and (7.130) for $\nu = 0.0$, 0.3, and 0.5 is shown in Fig. 7.20.

References

1. Öchsner A (2014) Elasto-plasticity of frame structure elements: modeling and simulation of rods and beams. Springer, Berlin
2. Öchsner A (2016) Continuum damage and fracture mechanics. Springer, Singapore

Appendix A
Mathematics

A.1 Greek Alphabet

The letters of the Greek alphabet are shown in Table A.1.

A.2 Frequently Used Constants

$$\pi = 3.14159 \,,$$
$$e = 2.71828 \,,$$
$$\sqrt{2} = 1.41421 \,,$$
$$\sqrt{3} = 1.73205 \,,$$
$$\sqrt{5} = 2.23606 \,,$$
$$\sqrt{e} = 1.64872 \,,$$
$$\sqrt{\pi} = 1.77245 \,.$$

A.3 Special Products

$$(x + y)^2 = x^2 + 2xy + y^2 \,, \tag{A.1}$$
$$(x - y)^2 = x^2 - 2xy + y^2 \,, \tag{A.2}$$

© The Editor(s) (if applicable) and The Author(s), under exclusive license to Springer
Nature Switzerland AG 2021
A. Öchsner, *Classical Beam Theories of Structural Mechanics*,
https://doi.org/10.1007/978-3-030-76035-9

Table A.1 The Greek alphabet

Name	Small letters	Capital letters
Alpha	α	A
Beta	β	B
Gamma	γ	Γ
Delta	δ	Δ
Epsilon	ϵ	E
Zeta	ζ	Z
Eta	η	H
Theta	θ, ϑ	Θ
Iota	ι	I
Kappa	κ	K
Lambda	λ	Λ
My	μ	M
Ny	ν	N
Xi	ξ	Ξ
Omikron	o	O
Pi	π	Π
Rho	ρ, ϱ	P
Sigma	σ	Σ
Tau	τ	T
Ypsilon	υ	Υ
Phi	ϕ, φ	Φ
Chi	χ	X
Psi	ψ	Ψ
Omega	ω	Ω

$$(x + y)^3 = x^3 + 3x^2 y + 3xy^2 + y^3, \tag{A.3}$$
$$(x - y)^3 = x^3 - 3x^2 y + 3xy^2 - y^3, \tag{A.4}$$
$$(x + y)^4 = x^4 + 4x^3 y + 6x^2 y^2 + 4xy^3 + y^4, \tag{A.5}$$
$$(x - y)^4 = x^4 - 4x^3 y + 6x^2 y^2 - 4xy^3 + y^4. \tag{A.6}$$

A.4 Trigonometric Functions

Definition on a right-angled triangle

The triangle ABC is in C right-angled and has edges of length a, b, c. The trigonometric functions of the angle α are defined in the following manner (see Fig. A.1):

Fig. A.1 Right-angled
triangle

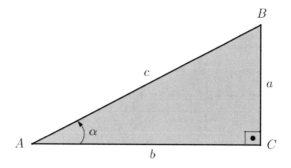

$$\text{sine of } \alpha = \sin \alpha = \frac{a}{c} = \frac{\text{opposite}}{\text{hypotenuse}}, \tag{A.7}$$

$$\text{cosine of } \alpha = \cos \alpha = \frac{b}{c} = \frac{\text{adjacent}}{\text{hypotenuse}}, \tag{A.8}$$

$$\text{tangent of } \alpha = \tan \alpha = \frac{a}{b} = \frac{\text{opposite}}{\text{adjacent}}, \tag{A.9}$$

$$\text{cotangent of } \alpha = \cot \alpha = \frac{b}{a} = \frac{\text{adjacent}}{\text{opposite}}, \tag{A.10}$$

$$\text{secant of } \alpha = \sec \alpha = \frac{c}{b} = \frac{\text{hypotenuse}}{\text{adjacent}}, \tag{A.11}$$

$$\text{cosecant of } \alpha = \csc \alpha = \frac{c}{a} = \frac{\text{hypotenuse}}{\text{opposite}}. \tag{A.12}$$

Addition formulae

$$\sin(\alpha \pm \beta) = \sin \alpha \cos \beta \pm \cos \alpha \sin \beta, \tag{A.13}$$
$$\cos(\alpha \pm \beta) = \cos \alpha \cos \beta \mp \sin \alpha \sin \beta, \tag{A.14}$$
$$\tan(\alpha \pm \beta) = \frac{\tan \alpha \pm \tan \beta}{1 \mp \tan \alpha \tan \beta}, \tag{A.15}$$
$$\cot(\alpha \pm \beta) = \frac{\cot \alpha \cot \beta \mp 1}{\cot \beta \pm \cot \beta}. \tag{A.16}$$

Identity formula

$$\sin^2 \alpha + \cos^2 \alpha = 1. \tag{A.17}$$

Analytic values for different angles (see Table A.2)

Table A.2 Analytical values of sine, cosine, tangent and cotangent for different angles

α in degree	α in radian	$\sin\alpha$	$\cos\alpha$	$\tan\alpha$	$\cot\alpha$
0°	0	0	1	0	$\pm\infty$
30°	$\frac{1}{6}\pi$	$\frac{1}{2}$	$\frac{\sqrt{3}}{2}$	$\frac{\sqrt{3}}{3}$	$\sqrt{3}$
45°	$\frac{1}{4}\pi$	$\frac{\sqrt{2}}{2}$	$\frac{\sqrt{2}}{2}$	1	1
60°	$\frac{1}{3}\pi$	$\frac{\sqrt{3}}{2}$	$\frac{1}{2}$	$\sqrt{3}$	$\frac{\sqrt{3}}{3}$
90°	$\frac{1}{2}\pi$	1	0	$\pm\infty$	0
120°	$\frac{2}{3}\pi$	$\frac{\sqrt{3}}{2}$	$-\frac{1}{2}$	$-\sqrt{3}$	$-\frac{\sqrt{3}}{3}$
135°	$\frac{3}{4}\pi$	$\frac{\sqrt{2}}{2}$	$-\frac{\sqrt{2}}{2}$	1	1
150°	$\frac{5}{6}\pi$	$\frac{1}{2}$	$-\frac{\sqrt{3}}{2}$	$-\frac{\sqrt{3}}{3}$	$-\sqrt{3}$
180°	π	0	-1	0	$\pm\infty$
210°	$\frac{7}{6}\pi$	$-\frac{1}{2}$	$-\frac{\sqrt{3}}{2}$	$\frac{\sqrt{3}}{3}$	$\sqrt{3}$
225°	$\frac{5}{4}\pi$	$-\frac{\sqrt{2}}{2}$	$-\frac{\sqrt{2}}{2}$	1	1
240°	$\frac{4}{3}\pi$	$-\frac{\sqrt{3}}{2}$	$-\frac{1}{2}$	$\sqrt{3}$	$\frac{\sqrt{3}}{3}$
270°	$\frac{3}{2}\pi$	-1	0	$\pm\infty$	0
300°	$\frac{5}{3}\pi$	$-\frac{\sqrt{3}}{2}$	$\frac{1}{2}$	$-\sqrt{3}$	$-\frac{\sqrt{3}}{3}$
315°	$\frac{7}{4}\pi$	$-\frac{\sqrt{2}}{2}$	$\frac{\sqrt{2}}{2}$	-1	-1
330°	$\frac{11}{6}\pi$	$-\frac{1}{2}$	$\frac{\sqrt{3}}{2}$	$-\frac{\sqrt{3}}{3}$	$-\sqrt{3}$
360°	2π	0	1	0	$\pm\infty$

Recursion formulae
Typical recursion formulae are summarized in Table A.3.

Table A.3 Recursion formulae for trigonometric functions

	$-\alpha$	$90°\pm\alpha$ $\frac{\pi}{2}\pm\alpha$	$180°\pm\alpha$ $\pi\pm\alpha$	$270°\pm\alpha$ $\frac{3\pi}{2}\pm\alpha$	$k(360°)\pm\alpha$ $2k\pi\pm\alpha$
sin	$-\sin\alpha$	$\cos\alpha$	$\mp\sin\alpha$	$-\cos\alpha$	$\pm\sin\alpha$
cos	$\cos\alpha$	$\mp\sin\alpha$	$-\cos\alpha$	$\pm\sin\alpha$	$\cos\alpha$
tan	$-\tan\alpha$	$\mp\cot\alpha$	$\pm\tan\alpha$	$\mp\cot\alpha$	$\pm\tan\alpha$
csc	$-\csc\alpha$	$\sec\alpha$	$\mp\csc\alpha$	$-\sec\alpha$	$\pm\csc\alpha$
sec	$\sec\alpha$	$\mp\csc\alpha$	$-\sec\alpha$	$\pm\csc\alpha$	$\sec\alpha$
cot	$-\cot\alpha$	$\mp\tan\alpha$	$\pm\cot\alpha$	$\mp\tan\alpha$	$\pm\cot\alpha$

A.5   Taylor's Series Expansion

A Taylor's series expansion of $f(x)$ with respect to x_0 is given by:

$$f(x) = f(x_0) + \left(\frac{\mathrm{d}f}{\mathrm{d}x}\right)_{x_0}(x - x_0) + \frac{1}{2!}\left(\frac{\mathrm{d}^2 f}{\mathrm{d}x^2}\right)_{x_0}(x - x_0)^2 + \cdots + \frac{1}{k!}\left(\frac{\mathrm{d}^k f}{\mathrm{d}x^k}\right)_{x_0}(x - x_0)^k .$$

(A.18)

The first order approximation takes the first two terms in the series and approximates the function as:

$$f(x) = f(x_0 + \mathrm{d}x) \approx f(x_0) + \left(\frac{\mathrm{d}f}{\mathrm{d}x}\right)_{x_0}(x - x_0) .$$

(A.19)

Recall from calculus that the derivative gives the slope of the tangent line at a given point and that the point-slope form is given by $f(x) - f(x_0) = m \times (x - x_0)$. Thus, we can conclude that the first order approximation gives us the equation of a straight line passing through the point $(x_0, f(x_0))$ with a slope of $m = f'(x_0) = (\mathrm{d}f/\mathrm{d}x)_{x_0}$, cf. Fig. A.2.

A.1   Example: Taylor's series expansion of a function

Calculate the first and second order Taylor's series approximation of the function $f(x) = x^2$ at the functional value $x = 2$ based on the value at $x_0 = 1$. The exact solution is $f(x = 2) = 2^2 = 4$.

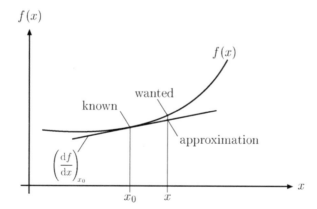

Fig. A.2 Approximation of a function $f(x)$ by a first order Taylor's series

A.1 Solution

First order approximation:

$$f(x = 2) \approx f(x_0 = 1) + 2x|_{x_0} \times (x - x_0) = 1 + 2 \times 1(2 - 1) = 3. \quad \text{(A.20)}$$

$$\text{rel. error} \left| \frac{4 - 3}{3} \right| = \left| \frac{1}{4} \right| = 0.25 = 25\%. \quad \text{(A.21)}$$

Second order approximation:

$$f(x = 2) \approx f(x_0 = 1) + 2x|_{x_0} \times (x - x_0) + \quad \text{(A.22)}$$

$$+ \left. \frac{1}{2} \times 2 \right|_{x_0} (x - x_0)^2 = 1 + 2 + 1 = 4 \ \checkmark \quad \text{(A.23)}$$

A.6 Analytical Geometry

• Point-slope form:

Given is a point (x_1, y_1) and a slope m:

$$y - y_1 = m(x - x_1), \quad \text{(A.24)}$$

$$y = (y_1 - mx_1) + m \times x. \quad \text{(A.25)}$$

• Slope-intercept form:

Given is a slope m and the y-intercept b:

$$y = b + m \times x. \quad \text{(A.26)}$$

• Two-point form:

Given are the points (x_1, y_1) and (x_2, y_2):

$$\frac{y - y_1}{x - x_1} = \frac{y_1 - y_2}{x_1 - x_2}, \quad \text{(A.27)}$$

$$y = \left(y_1 - \frac{y_1 - y_2}{x_1 - x_2} x_1 \right) + \frac{y_1 - y_2}{x_1 - x_2} \times x. \quad \text{(A.28)}$$

Appendix B
Mechanics

B.1 Centroids

The coordinates (y_c, z_c) of the centroid c of the plane surface shown in Fig. B.1 can be expressed as

$$z_c = \frac{\int z \, dA}{\int dA}, \tag{B.1}$$

$$y_c = \frac{\int y \, dA}{\int dA}, \tag{B.2}$$

where the integrals $\int z \, dA$ and $\int y \, dA$ are known as the first moments of area[1]. In the case of surfaces composed of n simple shapes, the integrals can be replaced by summations to obtain:

$$z_c = \frac{\sum\limits_{i=1}^{n} z_i A_i}{\sum\limits_{i=1}^{n} A_i}, \tag{B.3}$$

$$y_c = \frac{\sum\limits_{i=1}^{n} y_i A_i}{\sum\limits_{i=1}^{n} A_i}. \tag{B.4}$$

---

[1] A better expression would be moment of surface since area means strictly speaking the measure of the size of the surface which is different to the surface itself.

© The Editor(s) (if applicable) and The Author(s), under exclusive license to Springer Nature Switzerland AG 2021
A. Öchsner, *Classical Beam Theories of Structural Mechanics*,
https://doi.org/10.1007/978-3-030-76035-9

If the surface is doubly symmetric about two orthogonal axes, the centroid lies at the intersection of those axes. If the surface is singly symmetric about one axis, then the centroid lies somewhere along that axis (the other coordinate must be calculated).

B.2 Second Moment of Area

The second moment of area[2] or the second area moment is a geometrical property of a surface which reflects how its area elements are distributed with regard to an arbitrary axis. The second moments of area for an arbitrary surface with respect to an arbitrary Cartesian coordinate system (see Fig. B.1) are generally defined as:

$$I_y = \int_A z^2 \mathrm{d}A \,, \tag{B.5}$$

$$I_z = \int_A y^2 \mathrm{d}A \,. \tag{B.6}$$

These quantities are normally used in the context of plane bending of symmetrical cross sections. For unsymmetrical bending, the product moment of area is additionally required:

$$I_{yz} = -\int_A yz\mathrm{d}A \,. \tag{B.7}$$

In case of a pure rotation of the x-y coordinate system around its origin (which is identical with the centroid) by an rotational angle $\alpha^{y-y'}$ (see Fig. B.2), the following transformation relations can be used to express the second moments of area in the rotated x'-y' coordinate system:

$$I_{y'} = \frac{I_y + I_z}{2} + \frac{I_y - I_z}{2}\cos\left(2\alpha^{y-y'}\right) + I_{yz}\sin\left(2\alpha^{y-y'}\right) \,, \tag{B.8}$$

$$I_{z'} = \frac{I_y + I_z}{2} - \frac{I_y - I_z}{2}\cos\left(2\alpha^{y-y'}\right) - I_{yz}\sin\left(2\alpha^{y-y'}\right) \,, \tag{B.9}$$

$$I_{y'z'} = -\frac{I_y - I_z}{2}\sin\left(2\alpha^{y-y'}\right) + I_{yz}\cos\left(2\alpha^{y-y'}\right) \,. \tag{B.10}$$

---

[2]The second moment of area is also called in the literature the second moment of inertia. However, the expression moment of inertia is in the context of properties of surfaces misleading since no mass or movement is involved.

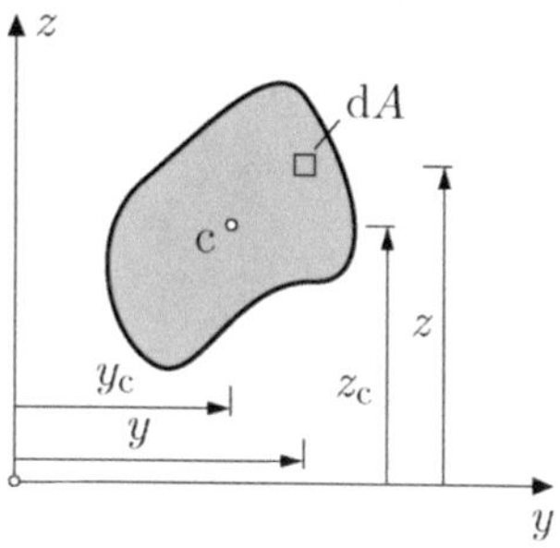

Fig. B.1 Plane surface with centroid c

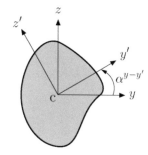

Fig. B.2 Rotation of a Cartesian x-y coordinate system around its origin (centroid) by an angle $\alpha^{y-y'}$

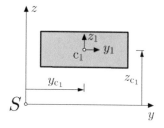

Fig. B.3 Configuration for the parallel-axis theorem

B.3 Parallel-Axis Theorem

The parallel-axis theorem gives the relationship between the second moment of area with respect to a centroidal axis (z_1, y_1) and the second moment of area with respect to any parallel axis[3] (z, y). For the rectangular shown in Fig. B.3, the relations can be expressed as:

$$I_y = I_{y_1} + z_{c_1}^2 \times A_1 , \tag{B.11}$$

$$I_z = I_{z_1} + y_{c_1}^2 \times A_1 , \tag{B.12}$$

$$I_{zy} = I_{z_1 y_1} - z_{c_1} y_{c_1} \times A_1 . \tag{B.13}$$

---

[3]This arbitrary axis can be for example the axis trough the common centroid c of a composed surface.

Index

Printed in the United States
by Baker & Taylor Publisher Services